THE FUTURE OF WARFARE

Environment, Geography, and the Future of Warfare

The Changing Global Environment and Its Implications for the U.S. Air Force

SHIRA EFRON, KURT KLEIN, AND RAPHAEL S. COHEN

PROJECT AIR FORCE

Prepared for the United States Air Force
Approved for public release; distribution unlimited

For more information on this publication, visit www.rand.org/t/RR2849z5

Library of Congress Cataloging-in-Publication Data is available for this publication.
ISBN: 978-1-9774-0299-8

Published by the RAND Corporation, Santa Monica, Calif.

Cover: Senior Airman Brittain Crolley, U.S. Air Force
Spine: combo1982/Getty Images, matejmo/Getty Images, StudioM1/Getty Images

Preface

Where will the next war occur? Who will fight in it? Why will it occur? How will it be fought? Researchers with RAND Project AIR FORCE's Strategy and Doctrine Program attempted to answer these questions about the future of warfare—specifically, those conflicts that will drive a U.S. and U.S. Air Force response—by examining the key geopolitical, economic, environmental, geographic, legal, informational, and military trends that will shape the contours of conflict between now and 2030. This report on environment, geography, and the future of warfare is one of a series that grew out of this effort. The other reports in the series are

- Raphael S. Cohen et al., *The Future of Warfare in 2030: Project Overview and Conclusions* (RR-2849/1-AF)
- Raphael S. Cohen, Eugeniu Han, and Ashley L. Rhoades, *Geopolitical Trends and the Future of Warfare: The Changing Global Environment and Its Implications for the U.S. Air Force* (RR-2849/2-AF)
- Forrest E. Morgan and Raphael S. Cohen, *Military Trends and the Future of Warfare: The Changing Global Environment and Its Implications for the U.S. Air Force* (RR-2849/3-AF)
- Howard J. Shatz and Nathan Chandler, *Global Economic Trends and the Future of Warfare: The Changing Global Environment and Its Implications for the U.S. Air Force* (RR-2849/4-AF)
- Bryan Frederick and Nathan Chandler, *Restraint and the Future of Warfare: The Changing Global Environment and Its Implications for the U.S. Air Force* (RR-2849/6-AF).

This volume examines six trends by asking four key questions for each trend. First, what does research say about how this variable shapes the conduct of warfare? Second, how has this variable historically shaped the conduct of warfare, especially in the post–Cold War era? Third, how might this variable be expected to change through 2030? And finally, but perhaps most importantly, how might this variable affect the future of warfare in this time frame, especially as it relates to the U.S. armed forces and the U.S. Air Force in particular? By answering these questions, it is hoped that this report will paint a picture of how environment and geography will contribute to shaping conflict over the next decade and beyond.

This research was sponsored by the Director of Strategy, Concepts and Assessments, Deputy Chief of Staff for Strategic Plans and Requirements, (AF/A5S). It is part of a larger study, entitled *The Future of Warfare*, that is intended to assist the Air Force in assessing trends in the future strategic environment. This report should be of value to the national security community and interested members of the general public, especially those with an interest in how global trends will affect the conduct of warfare. Comments are welcome and should be sent to the authors: Shira Efron, Kurt Klein, or Raphael S. Cohen (who is also the project leader). Research was completed in October 2018.

RAND Project AIR FORCE

RAND Project AIR FORCE (PAF), a division of the RAND Corporation, is the U.S. Air Force's federally funded research and development center for studies and analyses. PAF provides the Air Force with independent analyses of policy alternatives affecting the development, employment, combat readiness, and support of current and future air, space, and cyber forces. Research is conducted in four programs: Force Modernization and Employment; Manpower, Personnel, and Training; Resource Management; and Strategy and Doctrine. The research reported here was prepared under contract FA7014-16-D-1000.

Additional information about PAF is available on our website: www.rand.org/paf.

This report documents work originally shared with the U.S. Air Force on September 2018. The draft report, issued on September 18, 2018, was reviewed by formal peer reviewers and U.S. Air Force subject-matter experts.

Contents

Figures and Tables

Figures

Tables

Summary

Climate and geography shape where and why conflicts occur and how actors fight. Climate might not be the sole driver of conflict, but conflicts are frequently associated with the climates of the regions where they are fought. Furthermore, climatic factors exacerbate tensions, especially where adverse social conditions are present. As part of a larger effort to characterize the future of warfare in 2030, this volume assesses the implications of six intertwined key climate and geographical trends:

1. the steady rise of global temperatures
2. the opening of the Arctic
3. sea level rise
4. more-frequent extreme weather events
5. growing water scarcity
6. urbanization and the development of megacities.

We divide our analysis of each trend into four basic questions:

1. What has previous research shown about how the trend might shape the conduct of warfare?
2. How has this trend evolved in recent history?
3. How might this trend evolve in the next ten to 15 years?
4. What are the implications of this trend for the future of warfare in 2030, and the U.S. Air Force in particular?

Our analysis shows that although these trends are unlikely on their own to lead to conflict, they are threat multipliers that could exacerbate existing problems and fuel instability around the world. This suggests that despite a focus on peer competitors in the 2018 *National Defense Strategy*, the U.S. Air Force (USAF) will face continued demand for counterterrorism and stability operations—and perhaps increasing demand for humanitarian assistance/disaster relief missions in the years to come. Most climatic phenomena are likely to affect all countries and developing countries in particular—including areas of prime strategic importance for the United States, such as the Middle East and North Africa and South Asia. Moreover, these trends are likely to shape myriad military activities, including planning, acquisitions, basing, training, and operations. For example, 26 USAF installations have already been affected by flooding.[1] A site where the sea currently encroaches is the Marshall Islands, where a USAF radar installation worth nearly $1 billion could be underwater by 2035.[2] Furthermore, training areas will face risks of flooding or coastal erosion and of storm damage to base infrastructure. Service members will need to cope with a variety of challenges, such as dangerous heat or infectious disease outbreaks. The development of megacities could produce conditions of failed governance and lawlessness, which could require U.S. intervention. In these dense urban environments—characterized by tall buildings, narrow streets, and subterranean spaces that protect adversaries—the advantages of air superiority (specifically, aerial surveillance and close air support) are undermined. Furthermore, the risks of collateral damage increase substantially. To prepare for and mitigate against the consequences of these climate and geographical trends, we propose the following policy recommendations to the USAF:

---

[1] Office of the Under Secretary of Defense for Acquisition, Technology, and Logistics, *Department of Defense Climate-Related Risk to DoD Infrastructure Initial Vulnerability Assessment Survey (SLVAS) Report*, Washington, D.C.: U.S. Department of Defense, January 2018.

[2] Curt D. Storlazzi et al., *The Impact of Sea-Level Rise and Climate Change on Department of Defense Installations on Atolls in the Pacific Ocean*, Washington, D.C.: U.S. Department of Defense, RC-2334, February 2018.

- Cooperate with other branches of the U.S. government and scientific community to adapt climate data and analysis for USAF-specific needs.
- Create the USAF equivalent of the U.S. Navy's Task Force Climate Change to anticipate, analyze, and address specific climate change risks to the USAF mission.
- Incorporate risks associated with a changing climate and the proliferation of megacities into planning and wargames, including scenarios in which such trends fuel civil conflicts that could draw the United States into new theaters of warfare.
- Develop USAF-specific strategies and capabilities to operate in the Arctic, if necessary, including reliable communications and intelligence, surveillance, and reconnaissance.
- Assess current USAF infrastructure and think through the alternatives of adaptation and protection for future climate change. If justified and proven cost-effective, protect existing infrastructure or relocate and repurpose such infrastructure. In addition, design new infrastructure that can sustain heat, extreme weather, and sea level rise.
- Develop a doctrine and specialized equipment to operate in dense urban and megacity environments, including both capabilities to handle big data quickly and more-precise and nonlethal weapons to minimize collateral damage.

Acknowledgments

This study would not have been possible without the help of many people. First and foremost, we would like to thank Brig Gen David Hicks, Col Linc Bonner, and Scott Wheeler of the Air Force A5S for sponsoring this project and guiding it along the way. We would also like to thank Paula Thornhill, Project AIR FORCE strategy doctrine program director, for her guidance and mentorship of this study. For their detailed feedback and thoughtful comments on this report, we also thank our formal peer reviewers: Francesco Femia, cofounder of the Center for Climate and Security, and senior RAND Corporation researchers Robert Lempert and Debra Knopman. The research team owes a special debt of gratitude to dozens of experts at RAND and externally, across the globe, who volunteered their time to give their perspectives on the future of warfare both within their region and globally. Human subjects protocol prevents us from thanking them by name, but we thank the following institutions for hosting our research visits: Interdisciplinary Center Herzliya; University of Jordan; Jordanian Army Forces; Royal Jordanian Air Force; MEMPSI–Centre for Strategic and International Studies; Strategic Intelligence Solutions; U.S. Embassy Amman, Jordan; and U.S. Embassy, United Arab Emirates.

Abbreviations

DoD	U.S. Department of Defense
EEZ	Exclusive Economic Zone
FT	feet
HA/DR	humanitarian assistance and disaster relief
IPCC	Intergovernmental Panel on Climate Change
ISR	intelligence, surveillance and reconnaissance
MENA	Middle East and North Africa
NSR	Northern Sea Route
PAF	Project AIR FORCE
RCP	Representative Concentration Pathway
SLR	sea level rise
USAF	United States Air Force

CHAPTER ONE

Geographical and Environmental Trends

> Climate change can be a driver of instability and the Department of Defense must pay attention to potential adverse impacts generated by this phenomenon.
>
> —Secretary of Defense James Mattis, January 2017[1]

Geography, climate, and national security are all tightly linked. *Geography* describes places and processes affecting places.[2] *Climate* affects how people interact with these places—where they live, what resources they exploit, and what they must do to survive. As we shall see, climate and geography shape where and why conflicts occur. Climate might not be the sole driver of conflict in some contexts (or even the primary one), but research has indicated links between climate and conflict.[3] More importantly, climate can exacerbate tensions, especially in places vulnerable to instability resulting from adverse social conditions.[4] Consequently, understanding how climate and geography will change over the next decade is crucial to understanding the future of warfare in

---

1 Andrew Revkin, "Trump's Defense Secretary Cites Climate Change as National Security Challenge," ProPublica, March 14, 2017.

2 Albert Jackman, "The Nature of Military Geography," *The Professional Geographer*, Vol. 14, 1962.

3 Solomon M. Hsiang, Marshall Burke, and Edward Miguel, "Quantifying the Influence of Climate on Human Conflict," *Science*, Vol. 341, No. 6151, 2013.

4 Jesse Ribot, "Vulnerability Does Not Fall from the Sky: Toward Multiscale, Pro-Poor Climate Policy," in Robin Mearns and Andrew Norton, eds., *Social Dimensions of Climate Change: Equity and Vulnerability in a Warming World*, Washington, D.C.: World Bank, 2010.

2030. In our analysis, we are agnostic on the question of whether any specific trend can be attributed to anthropogenic climate change or is the result of natural climate variation. However, we do occasionally refer to models that link climate change with man-made carbon emissions (e.g., the rise of temperatures under four Representative Concentration Pathways [RCPs] in Trend 1) because, from a risk management perspective, the U.S. Air Force (USAF) should account for high-risk scenarios, which are associated with man-made climate variability. Our analysis focuses on the possible ramifications of six climate and geographical trends: the steady rise of global temperatures, the opening of the Arctic, sea level rise (SLR), more-frequent and more-extreme weather events, growing water scarcity, and the development of megacities. These trends could have ramifications for who, where, and how the United States fights.

Methodology, Limitations, and Bounding the Scope of Analysis

We selected the aforementioned six trends as being most likely to interest the USAF. As with the other themes explored in the Future of Warfare series—geopolitical, military, space and nuclear, cyber, restraint, and global economic trends—we divide the analysis of each trend into four basic questions. First, what does previous research say about how the variable could shape the conduct of warfare? Second, how has this variable evolved in recent years? Third, how might this variable evolve through 2030? And finally, what are the implications of this variable for the future of warfare, and where possible, for the USAF in particular? Our analysis of each of these four questions was based on reviewing relevant literature and speaking with subject-matter experts at the RAND Corporation and externally. We created unique maps for this report using publicly available data and ArcGIS.

Before delving into the analysis, we note some methodological caveats. First, trends covered in this volume were chosen by RAND experts in consultation with the USAF. These are the key climatic and geographical trends that subject-matter experts associated with the most-pressing implications for U.S. national security and the U.S. military in the near-medium-term future. These six trends are not exhaustive and more research is needed to understand how each variable in this area could affect warfare. Second, although we can speak broadly about trends and anticipate that the frequency and severity of climate events will increase overall, experts in many instances still cannot project the exact location, timing, and magnitude of such events and how they might cascade.[5] To a large extent, the consequences of climatic events and of larger trends depend on the capacity of institutions, governments, and regions to cope with such challenges. Thus, similar phenomena can have very different outcomes depending on their contexts, on the prevailing socioeconomic and political conditions, and on the interventions adopted. Where applicable, we state that our projections assume present trends will continue and that no meaningful intervention will be adopted. On the other hand, some trends discussed in this volume are understood better and with a higher degree of confidence among the scientific community. For example, Trend 1 (about rising temperatures) discusses two issues: the impact of high temperatures on people's ability to work outdoors in such areas as the Middle East—which scientists understand with high confidence—and the effects of rising temperatures on state stability, which is not as well understood. To differentiate between our levels of understandings of phenomena, we used the Intergovernmental Panel on Climate Change's (IPCC's) Guidance on Consistent Treatment of Uncertainties, which uses the terms *confidence* and *likelihood*.[6]

Although the most-dramatic effects of climate on conflict will take place in the more distant future (with assessments focusing on the years 2050 and 2100), climate has already contributed to conflict (e.g., in Syria, Kenya, and other places we discuss in Chapter Six) and is

---

[5] John D. Steinbruner, Paul C. Stern, and Jo L. Husbands, eds., *Climate and Social Stress: Implications for Security Analysis*, Washington, D.C.: Committee on Assessing the Impacts of Climate Change on Social and Political Stresses, Board on Environmental Change and Society, Division of Behavioral and Social Sciences and Education, National Research Council, 2013.

[6] IPCC, "Guidance Note for Lead Authors of the IPCC Fifth Assessment Report on Consistent Treatment of Uncertainties," IPCC Cross-Working Group Meeting on Consistent Treatment of Uncertainties, Jasper Ridge, Calif., July 6–7, 2010.

likely to shape conflict in 2030 in some places. We attempt to analyze the trends separately and artificially separate the discussion on climatic effects into five of our chosen trends (the steady rise of global temperatures, opening of the Arctic, SLR, more-frequent and more-extreme weather events, and growing water scarcity), but we acknowledge that significant synergies exist among them. In practice, they all are driven by well-documented patterns in warming of surface temperatures and oceans and are interdependent as a result. Moreover, they also relate to our sixth trend (expansion and proliferation of megacities) because changes to the environment incite migration, and large cities are most often the destination for people who are forced to move. Finally, megacities are often located in coastal areas, which are most vulnerable to SLR. The challenges for the USAF and the populations that experience the consequences that arise from each trend are deeply intertwined with challenges stemming from the other trends.

CHAPTER TWO

Trend 1: Rising Temperatures

Although there is significant variability from year to year and across regions, one of the central effects of changes in climate over the past century has been the steady rise in global average surface temperature, both on land and sea. Although warming has not been uniform across the globe, the upward trend in the globally averaged temperature shows that more areas are warming than cooling. Rising temperatures and the increase in the number of heat waves, superimposed on a highly variable system, have several direct and indirect implications for national security and warfighting.

Context: Higher Temperatures Pose Risks to Human Health; Reduce Water, Food, and Energy Security; and Increase the Likelihood of Conflict

Global temperatures have risen steadily over the past three to four decades at an accelerating pace.[1] Extreme heat is dangerous for human health; it is the deadliest natural disaster in the United States, killing more people on average than other types of extreme weather events and natural disasters combined.[2] Heat waves have also had severe health

1 National Centers for Environmental Information, "Global Climate Report—Annual 2016," webpage, undated.

2 Center for Climate and Energy Solutions, "Heat Waves and Climate Change," webpage, undated.

effects in the Middle East,[3] Africa, Europe, and South Asia.[4] High temperatures prolong droughts and make them more extreme, which could lead to further increases in temperatures as the sun's energy is turned to heating air and land surface instead of evaporating water.[5] Hot, dry conditions also exacerbate the risk of wildfire.[6] Moreover, higher temperatures increase electricity demands,[7] and at the same time they hinder the transmission capacity of power lines, making electricity supply less reliable.[8]

A meta-analysis of dozens of articles in a variety of fields—archaeology, criminology, economics, geography, history, political science, and psychology—found that increases in temperature and extreme rainfall are associated across different regions with increases in conflict both between groups and between individuals.[9] Heat induces aggression while lowering human productivity, and abnormally hot months in the United States have been associated with higher crime rates in cities.[10] Aggression and violence also increase as distance from the equator shortens and temperatures rise.[11] Warmer weather has

---

[3] Hugh Naylor, "An Epic Middle East Heat Wave Could Be Global Warming's Hellish Curtain-Raiser," *Washington Post*, August 10, 2016.

[4] Jeff Masters, "Historic Heat Wave Sweeps Asia, the Middle East and Europe," *Weather Underground*, June 6, 2017.

[5] Center for Climate and Energy Solutions, undated.

[6] Center for Climate and Energy Solutions, undated.

[7] Jan Dell, Susan Tierney, Guido Franco, Richard G. Newell, Rich Richels, John Weyant, and Thomas J. Wilbanks, "Energy Supply and Use," in Jerry M. Melillo, Terese C. Richmond, and Gary W. Yohe, eds., *Climate Change Impacts in the United States: The Third National Climate Assessment*, Washington, D.C.: U.S. Global Change Research Program, 2014, pp. 113–129.

[8] U.S. Department of Energy, *U.S. Energy Sector Vulnerabilities to Climate Change and Extreme Weather*, Washington, D.C., July 2013.

[9] Hsiang, Burke, and Miguel, 2013.

[10] Tamma A. Carleton and Solomon M. Hsiang, "Social and Economic Impacts of Climate." *Science*, 2016.

[11] Paul A. M. Van Lange, Maria I. Rinderu, and Brad J. Bushman, "Aggression and Violence Around the World: A Model of Climate, Aggression, and Self-Control in Humans (CLASH)," *Behavioral and Brain Sciences*, Vol. 40, 2017.

shown to increase insurgency in India,[12] land invasions in Brazil,[13] and civil war intensity in Somalia.[14] Even when controlling for other factors, the relationship between heat and violence is linear, with violence rising approximately 11 percent per standard deviation in temperature.[15] Researchers also found that heat waves and droughts significantly increase the risk of armed conflict in ethnically fractionalized countries.[16]

There are several theories about why higher temperatures are associated with increases in conflict. First, higher temperatures reduce economic productivity and thus increase the value of engaging in conflict.[17] Reduced economic productivity also could curtail the ability of government institutions.[18] Droughts disrupt economies and displace populations, widening socioeconomic gaps and fueling economic grievances that cause conflict.[19] Finally, rising temperatures increase

---

12 Thiemo Fetzer, *Can Workfare Programs Moderate Violence? Evidence from India*, London: London School of Economics, working paper, June 2, 2014.

13 F. Daniel Hidalgo, Suresh Naidu, Simeon Nichter, and Neal Richardson, "Economic Determinants of Land Invasions," *The Review of Economics and Statistics*, Vol. 92, No. 3, 2010.

14 Jean-Francois Maystadt, Olivier Ecker, and Athur Mabiso, *Extreme Weather and Civil War in Somalia: Does Drought Fuel Conflict Through Livestock Price Shocks?* Washington, D.C.: International Food Policy Research Institute, IFPRI Discussion Paper 01243, 2013.

15 Carleton and Hsiang, 2016.

16 Carl-Friedrich Schleussner, Jonathan F. Donges, Reik V. Donner, and Hans Joachim Schellnhuber, "Armed-Conflict Risks Enhanced by Climate-Related Disasters in Ethnically Fractionalized Countries," *Proceedings of the National Academy of Sciences*, Vol. 113, No. 33, 2016, pp. 9216–9221.

17 For example, see Mariaflavia Harari and Eliana La Ferrara, "Conflict, Climate and Cells: A Disaggregated Analysis," *Review of Economics and Statistics*, Vol. 100, No. 4, January 2013; and Joshua Graff Zivin and Matthew Neidell, "Temperature and the Allocation of Time: Implications for Climate Change," *Journal of Labor Economics*, Vol. 32, No. 1, 2014.

18 Paul J. Burke and Andrew Leigh, "Do Output Contractions Trigger Democratic Change?" *American Economic Journal: Macroeconomics*, Vol. 2, No. 4, 2010; Eric Chaney, "Revolt on the Nile: Economic Shocks, Religion and Political Influence," *Topics in Middle Eastern and North African Economies*, Vol. 13, 2011.

19 Craig A. Anderson, Kathryn B. Anderson, Nancy Dorr, Kristina M. DeNeve, and Mindy Flanagan, "Temperature and Aggression," *Advances in Experimental Social Psychology*, Vol. 32,

the incidence of pests and the spread of lethal disease.[20] Several diseases are sensitive to extreme heat, such as ebola, malaria, Dengue fever, West Nile virus, cholera, and Lyme disease.[21]

Historical Trend: Temperatures Have Been Steadily Rising

Since 1970, global surface temperature rose at an average rate of about 0.3° Fahrenheit (0.17° Celsius) per decade, more than twice as fast as the 0.12° Fahrenheit (0.07° Celsius) per decade average recorded since 1880.[22] Earth's global surface temperatures in 2017 ranked as the second warmest since 1880 (after 2016), with global temperatures averaging 1.62° Fahrenheit (0.90° Celsius) warmer than the 1951–1980 mean.[23] Since 2010, scientists have observed the six warmest years on record, with four of those being set consecutively (2014–2017). From 1900 to 1980, a new temperature record was set on average every 13.5 years; since 1981, that rate has increased to every three years.[24] In addition to a steady rise in average temperatures, heat waves have become more frequent and longer.[25]

Specific instances of extreme temperatures have been felt across the Middle East and South Asia in recent years. In 2017, Ahvaz, a city in southwest Iran, suffered the highest temperature ever recorded

---

2000; Brian Jacob, Lars Lefgren, and Enrico Moretti, "The Dynamics of Criminal Behavior Evidence from Weather Shocks," *Journal of Human Resources*, Vol. 42, No. 3, 2007.

[20] Paul Parham, "Hard Evidence: Will Climate Change Affect the Spread of Tropical Diseases?" *The Conversation*, February 17, 2015.

[21] Anthony J. McMichael and Andrew Haines, "Global Climate Change: The Potential Effects on Health," *British Medical Journal Clinical Research*, Vol. 315, No. 7111, 1997, p. 805.

[22] Rebecca Lindsey and LuAnn Dahlman, "Climate Change: Global Temperature," climate.gov, September 11, 2017.

[23] National Aeronautics and Space Administration Goddard Institute for Space Studies, "Long-Term Warming Trend Continued in 2017: NASA, NOAA," webpage, January 18, 2018.

[24] National Oceanic and Atmospheric Administration, National Centers for Environmental Information, 2016.

[25] Center for Climate and Energy Solutions, undated.

in the country and the highest June temperature in mainland Asia—128.7° Fahrenheit.[26] This event is second only to when Mitribah, Kuwait, recorded a record temperature of 129.2° Fahrenheit in July 2016.[27] Similarly, May 2017 brought an unprecedented heatwave to several countries in the Middle East: Turbat, a city in west Pakistan, reached 128.3° Fahrenheit.[28]

Future Projection: More Hot Days, Heat Waves, and Risks to Human Health

Models predict a global increase in the number of hot days and heat waves.[29] Regardless of policy choices concerning emissions, global surface air temperature is expected to be 0.9° to 1.0° Fahrenheit warmer than the 1971–1999 average by 2030.[30] Figure 2.1 illustrates the rise of temperatures under two Representative Concentration Pathways (RCPs), which are possible greenhouse gas emission scenarios resulting from a variety of variables, such as use of fossil fuels.

Rising temperatures in coming decades are likely to affect turbulent regions severely, making hot and dry places even hotter and drier. Effects are likely to be most pronounced in the Middle East around the Persian Gulf and Red Sea, and in South Asia in the Indus and Ganges river valleys (Pakistan, Nepal, India, Bangladesh, and Sri Lanka).[31] Experts agree with a high degree of confidence that extreme heat will

---

[26] Jason Samenow, "Iranian City Soars to Record 129 Degrees: Near Hottest on Earth in Modern Measurements," *Washington Post*, June 29, 2017.

[27] Jason Samenow, "Two Middle East Locations Hit 129 Degrees, Hottest Ever in Eastern Hemisphere, Maybe the World," *Washington Post*, July 22, 2016.

[28] Masters, 2017.

[29] IPCC Working Group II, *Climate Change 2014: Impacts, Adaptation and Vulnerability*, Part A: *Global and Sectoral Aspects*, New York: Cambridge University Press, 2014.

[30] David Herring, "Climate Change: Global Temperature Projections," climate.gov, March 6, 2012.

[31] Eun-Soon Im, Jeremy S. Pal, and Elfatih A. B. Eltahir, "Deadly Heat Waves Projected in the Densely Populated Agricultural Regions of South Asia," *Science Advances*, Vol. 3, No. 8, 2017.

Figure 2.1
Global Average Temperature Change: 2000–2100

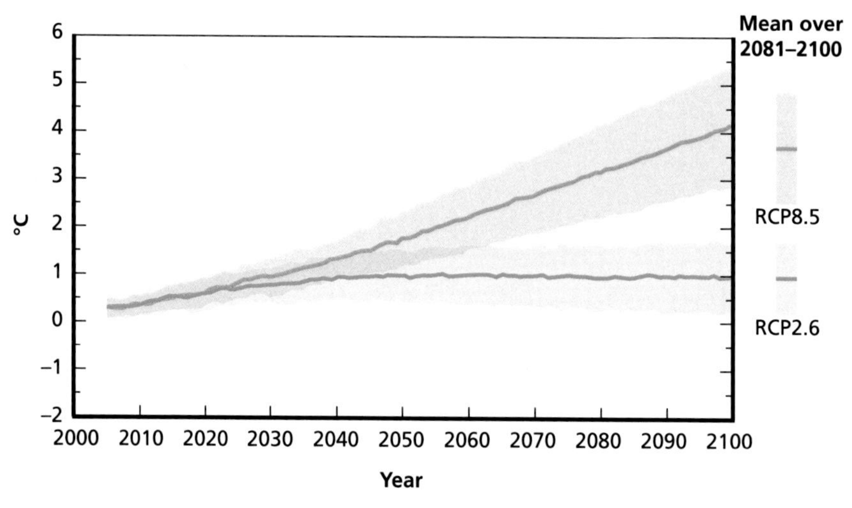

SOURCE: Rajendra K. Pachauri et al., *Climate Change 2014 Synthesis Report*, Geneva: Intergovernmental Panel on Climate Change, 2014.
NOTE: RCP2.6 is a mitigation scenario, and RCP8.5 is a very high–emission scenario reflecting no effort to contain emissions.

make parts of the Persian Gulf and its surroundings unlivable, potentially displacing populations living there,[32] and likely increasing the risk of infectious disease outbreak.[33] As already mentioned, higher temperatures also increase the risk of conflict, often substantially. One study suggested that if current trends continue, such a rise in temperature could increase the rate of conflicts by 50 percent in various regions of the world.[34]

---

[32] Christoph Schär, "Climate Extremes: The Worst Heat Waves to Come," *Nature Climate Change*, Vol. 6, No. 2, 2016.

[33] McMichael and Haines, 1997.

[34] Hsiang, Burke, and Miguel, 2013.

Implications for the U.S. Air Force and the Future of Warfare

Rising temperatures will have serious direct and indirect implications for the future of warfare and the USAF in coming decades. First, if mitigation measures are not implemented, high temperatures will jeopardize the health of service members. Increased heat also could affect training by lowering the performance of troops and possibly reducing the number of days safe for training and operation in hot areas. Moreover, considering that the annual number of hot days is predicted to increase globally, existing airbase infrastructure in especially warm places could be insufficient to support operations if concrete slabs at airports buckle and asphalt melts.[35]

Higher temperatures also affect aircraft. Arizona is home to seven military bases, including three USAF bases; in the summer of 2017, flights were regularly canceled or delayed in Phoenix, Arizona, when temperatures surpassed 120°.[36] By the same token, high summer temperatures at Ali Al Salem Air Base in Kuwait shut down some of the civilian aircraft electronic systems and regularly grounded flights, affecting USAF operations.[37] High temperatures also reduce air density, in turn reducing lift and constraining payload weight.[38] As the annual number of extremely hot days increases in future years, these impediments will become more frequent. U.S. officials project this trend will further complicate military aviation operations in the region over the long term.[39]

---

35 Alexandra Sims, "India's Roads Melt as Record-Breaking Heat Wave Continues," *Independent*, May 23, 2016.

36 Robert Ferris, "There's a Scientific Reason Why Hot Weather Has Grounded Planes at Phoenix Airport," CNBC, June 20, 2017.

37 "U.S. Military Bases in Kuwait," MilitaryBases.com, undated.

38 Ethan D. Coffel, Terence R. Thompson, and Radley M. Horton, "The Impacts of Rising Temperatures on Aircraft Takeoff Performance," *Climatic Change*, Vol. 144, No. 2, 2017.

39 Interviews with U.S. State Department and USAF officials, Abu Dhabi, United Arab Emirates, May 15, 2018.

Given forecasts for unlivable conditions in the Persian Gulf, more heat-related problems are anticipated in this turbulent part of the world where the U.S. military, and the USAF in particular, have substantial presence. Assuming the Middle East and South Asia continue to require U.S. military commitment, the U.S. Department of Defense (DoD) and USAF will need to invest heavily in increasing the heat-resilience of airbases and aircraft, modifying training methods and ranges, and finding medical solutions to heat-related problems. Table 2.1 provides a list of U.S. military installations and facilities in the Middle East where temperatures above 110° Fahrenheit have been recorded in recent years.

Even in areas where the military does not regularly operate, rising temperatures and their second-order effects (SLR, growing water and food scarcity, and increases in the spread of disease) could provoke more threats to political stability in foreign countries, conflicts, and crises. These events could incite responses from DoD and the USAF likely in the form of counterterrorism and humanitarian assistance and disaster relief (HA/DR).

Table 2.1
U.S. Military Installations and Facilities in the Middle East with Recorded Temperatures of More Than 110° F

Country	Installation/Facility	Main Use	Record Temperature
Bahrain (three sites; more than 7,000 U.S. military personnel)	Naval Support Activity	Naval Forces U.S. Central Command; headquarters of 5th Fleet	114° F
	Shaikh Isa Airbase	USAF has several warehouses of prepositioned equipment and supplies; previously, U.S. service members were deployed there; 12,467-feet (FT) runway	
	Muharraq Airfield, run by U.S. Navy	Serves as last stop for most U.S. troops headed to Afghanistan; two runways (13,000 FT and 8,300 FT)	
Egypt (Cairo)	Naval Medical Research Unit Three	Largest DoD overseas laboratory; conducts infectious disease research and prevention	116° F
Iraq (Anbar)	Al Asad Air Base	Task Force Lion (primary U.S. presence in Iraq) operates out of the base; two 13,124-FT runways	113° F
Israel (Negev)	Dimona Radar Facility	Used for the AN/TPY-2 radar system, a type of X-band radar	114° F
Kuwait (four sites)	Ali Al Salem Air Base	Hosts the 386th Air Expeditionary Wing, which operates C-17 and C-130 cargo aircraft; also serves as a base station for troops; two 9,805-FT runways	129° F
	Camp Arifjan	U.S. headquarters in Kuwait	
	Camp Buehring	Has precision approach radar capability for low visibility landings; 5,215-FT runway	
	Camp Patriot	U.S. Army facility	

Table 2.1—Continued

Country	Installation/Facility	Main Use	Record Temperature
Oman	U.S. uses six Omani bases per an access agreement	U.S. relied on air bases in early stages of Operation Enduring Freedom; Oman has allowed 5,000 overflights, 600 landings, and 80 port calls annually	122° F
Qatar (two sites; hosts 10,000 service members)	Al Udeid Air Base	Largest U.S. base in the Middle East; home to several commands, including U.S. combined Air Operations Center and U.S. Air Forces Central Command; used heavily in Operation Inherent Resolve; two runways of 12,000 FT	122° F
	Camp As Sayliyah	Army-operated facility; features many climate-controlled warehouse units	
United Arab Emirates (three sites; 5,000 U.S. military personnel)	Al Dhafra Air Base	Hosts the 380th Air Expeditionary Wing, including such aircraft as F-22 Raptor, KC-10, and RQ-4 Global Hawk; some 4,000 U.S. personnel believed stationed there; hosts the training facility—Joint Air Warfare Center	124° F
	Port of Jebel Ali	Busiest U.S. Navy port of call; capable of docking U.S. aircraft carriers	
	Fujairah Naval Base	Strategically located to serve as a logistical "land link" in case the Strait of Hormuz is closed	

SOURCE: Matthew Wallin, "U.S. Military Bases and Facilities in the Middle East," Washington, D.C.: American Security Project, fact sheet, June 2018.

CHAPTER THREE

Trend 2: Opening of the Arctic

Jeremy Mathis, director of the Arctic Research Program of the National Oceanic and Atmospheric Administration, stated in 2017 that "the Arctic is going through the most unprecedented transition in human history."[1] The pace of temperature increase is occurring twice as fast in the Arctic as in the rest of the world.[2] Since 2006, the centuries-long dream to navigate the Northwest Passage exclusively by water has been achievable almost every summer.[3] Arctic ice melt will prompt two primary changes in the security environment: expanded maritime access and increased economic activity as previously unreachable resources become extractable.

Context: Geopolitical Competition or Increased Cooperation?

The possibility of warfare in the far north depends on the interests of nations involved—most notably Russia and China, from the U.S. perspective. Russia's Arctic policies exist in a duality between compe-

[1] Chris Mooney, "Warming of the Arctic Is 'Unprecedented over the Last 1,500 Years,' Scientists Say," *Washington Post*, December 12, 2017.

[2] Mooney, 2017.

[3] Tom Di Liberto, "Northwest Passage Clear of Ice Again in 2016," climate.gov, September 16, 2016.

tition and cooperation.[4] On one hand, Russia has ambitious development plans in the region, developing energy and transportation infrastructure and building security infrastructure. On the other hand, the region offers many forums for Arctic nations to cooperate with each other,[5] and Russia benefits from this, publicly calling for the Arctic to remain a "zone of peace, stability and mutual cooperation."[6] Russia's willingness to choose competition (or even conflict) over cooperation is a function of how it perceives the costs and benefits of each option.[7] A more open Arctic will decrease the costs of operating in the area and increase the potential benefits as valuable resources and territory become increasingly accessible. Russia does not have the same level of Arctic capability that it did during the Cold War, but its maritime capabilities are significantly greater than that of the United States. For example, Russia has a fleet of 41 icebreakers; the United States has two.[8]

China is also playing a growing role in the Arctic. Viewing itself as a near Arctic state, China published its own Arctic policy, stating that it wishes to develop shipping routes, explore energy and fishing resources, and encourage Arctic tourism.[9] There is speculation that Chinese interest in the area is potentially hostile,[10] but academic

---

[4] Heather A. Conley and Caroline Rohloff, *The New Ice Curtain: Russia's Strategic Reach to the Arctic*, New York: Center for Strategic and International Studies, 2015, p. 112.

[5] Stephanie Pezard, Abbie Tingstad, Kristin Van Abel, and Scott Stephenson, *Maintaining Arctic Cooperation with Russia: Planning for Regional Change in the Far North*, Santa Monica, Calif.: RAND Corporation, RR-1731-RC, 2017, p. 21.

[6] John Thompson, "Putin Plays Mr. Nice Guy at Russian Arctic Forum," *Arctic Deeply*, March 30, 2017.

[7] Pezard et al., 2017, pp. 4–5.

[8] David Hambling, "Does the U.S. Stand a Chance Against Russia's Icebreakers?" *Popular Mechanics*, April 4, 2018; Pezard et al., 2017, pp. 14–15.

[9] Charlotte Gao, "China Issues Its Arctic Policy," *The Diplomat*, January 26, 2018.

[10] Olga V. Alexeeva and Frédéric Lasserre, "China and the Arctic," *Arctic Yearbook 2012*, Iceland: Arctic Portal, 2012, p. 80; Tom O'Connor, "China and Russia May Take Over the Top of the World with New 'Polar Silk Road' Through the Arctic," *Newsweek*, January 26, 2018.

research argues that China has more to gain by cooperating to develop economic interests and buying energy from Arctic nations.[11]

Historical Trend: Cooperation in the Arctic Despite Tension Elsewhere

An ice-free Arctic is a phenomenon never seen before and the most significant opening of untouched territory in the modern era, so historical trends might not be relevant.[12] But there are some noteworthy lessons to learn from some of the great powers' past behavior in the region.

Historically, the Arctic provided opportunities for the region's nations to cooperate. Even during the Cold War, the Soviet Union agreed to a framework to manage fisheries in the Barents Sea with Norway.[13] The Soviet Union, the United States, Canada, Denmark, and Norway signed the Agreement on the Conservation of Polar Bears in 1973.[14] In the 1987 Murmansk Initiative, Mikhail Gorbachev declared the Arctic a "zone of peace" and called for restricted military activity in the region; cooperation in resource development, indigenous people's affairs, and scientific research; and enhanced protection of the environment.[15]

The historical trend of cooperation continued after the end of the Cold War. Table 3.1 shows several examples of cooperation among Arctic nations in such fields as security (e.g., Arctic Security Forces Roundtable), science (e.g., the International Arctic Science Committee), and sovereignty claims (e.g., United Nations Convention on the

---

[11] Alexeeva and Lasserre, 2012, pp. 80–86.

[12] Jochen Knies, Patricia Cabedo-Sanz, Simon T. Belt, Soma Baranwal, Susanne Fietz, and Antoni Rosell-Melé, "The Emergence of Modern Sea Ice Cover in the Arctic Ocean," *Nature Communications*, Vol. 5, 2014.

[13] Pezard et al., 2017, p. 23.

[14] "Agreement on Conservation of Polar Bears," signed by the governments of Canada, Denmark, Norway, the Union of Soviet Socialist Republics and the United States of America, Oslo, November 15, 1973.

[15] Kristian Åtland, "Mikhail Gorbachev, the Murmansk Initiative, and the Desecuritization of Interstate Relations in the Arctic," *Cooperation and Conflict*, Vol. 43, No. 3, 2008.

Table 3.1
Arctic Nations Have a Strong History of Partnerships and Collaborations

Cooperation Examples[a]	United States	Russia	Canada	Denmark	Iceland	Norway	Sweden	Finland
Arctic Council	√	√	√	√	√	√	√	√
International Maritime Organization	√	√	√	√	√	√	√	√
International Arctic Science Committee	√	√	√	√	√	√	√	√
Arctic Security Forces Roundtable	√	√[b]	√	√	√	√	√	√
North Atlantic Treaty Organization	√		√	√	√	√		
Treaties on environment[c]	√	√	√	√	√	√	√	√
Arctic Council Search and Rescue Agreement	√	√	√	√	√	√	√	√
United Nations Convention on the Law of the Sea		√	√	√	√	√	√	√
Exercise Cold Response (2006, 2008, 2010, 2012, 2014)	√		√	√		√	√	√
Operation NANOOK (2007–2015)	√		√	√				
SAREX Greenland Sea (2012, 2013)		√	√	√	√	√		

SOURCE: Reprinted from Pezard et al., 2017, p. 21.

NOTE: Checkmarks indicate a country's participation.

[a] Collaborative bodies, treaties, and events.

[b] Russia not included in 2014.

[c] Includes the Agreement on the Conservation of Polar Bears and the Agreement on Cooperation on Marine Oil Pollution, Preparedness and Response in the Arctic.

Law of the Sea).[16] Notably, even during the February 2014 crisis over the Russian annexation of Crimea, the eight Arctic nations established the Arctic Economic Council and the Arctic Coast Guard Forum.[17]

However, this trend of cooperation might not endure. For example, after the United States and key allies sanctioned Russia over its revanchist military actions in Ukraine, Russia stopped participating in the Arctic Security Forces Roundtable.[18] Russia also has made aggressive moves in Arctic territory recently. Notably, Moscow announced that it would reopen 50 previously closed Arctic military bases from the Soviet era.[19] In 2007, Artur Chilingarov, a notable Russian oceanographer, planted a Russian flag on the seabed of the North Pole during a research expedition.[20] In March 2015, the Russian military conducted an exercise in the Arctic using more than 45,000 troops, 15 submarines, and 41 warships.[21]

Russia sees the Arctic as a source of national pride and wealth. For example, Prime Minister Dmitry Rogozin referred to the region as "Russia's Mecca,"[22] and roughly 20 percent of its gross domestic product comes from this area.[23] The Arctic draws Russians to it largely because of its economic significance and its status as a premise of patriotic sentiments.

Some analysts discount the perception that Russia is "militarizing the Arctic"; rather, they posit, it is "militarizing in the Arctic."[24] When

---

16 Pezard et al., 2017, pp. 20–22.

17 Pezard et al., 2017, pp. 2–4.

18 Michael Byers, "Crises and International Cooperation: An Arctic Case Study," *International Relations*, Vol. 31, No. 4, 2017, p. 385.

19 Conley and Rohloff, 2015, p. IX.

20 Pezard et al., 2017, p. 18. Scholars observed that Chilingarov likely planted the Russian flag on the Arctic seabed to further his own political career rather than as part of a Russian attempt to advance its territorial claims. While on the privately funded expedition, he was a member of the Russian Duma and was in the middle of an election.

21 Conley and Rohloff, 2015, p. IX.

22 Conley and Rohloff, 2015, pp. VIII–IX.

23 Lincoln Edson Flake, "Russia's Security Intentions in a Melting Arctic," *Military and Strategic Affairs*, Vol. 6, No. 1, 2014, p. 105.

24 Pezard et al., 2017, p. 15.

viewed more broadly, Russia is raising its military presence on its entire periphery, which includes the Arctic and could explain Russia's increased military activity there.[25] Russian security and economic interests (as well as internal politics) also could spill over to the Arctic, but the opening of the Arctic likely did not incentivize Russia to take a more aggressive posture. Indeed, the opening of the Arctic permits an increased military presence and could make conflict more possible; Russia might view this trend as a threat because it could cause other nations to enter its Arctic sphere of influence. As mentioned previously, Russia has a much greater maritime capability in the Arctic than the United States but is still far from establishing the military presence it had during the Cold War,[26] and the costs of operating in the harsh environment remain relatively high. Overall, Russia enjoys the status quo and likely does not want to engage in costly armed conflict in the region.[27]

In contrast, China has little history in the Arctic and developed its presence there only recently. As its economy has grown, China has become more interested in the Arctic's untapped resources and potential trade routes. Since entering the region, China has pursued joint ventures with Russian energy companies and invested in mines in Greenland.[28] So far, China's efforts to court Arctic nations has paid off; in 2013, it was granted observer status on the Arctic Council.[29]

Future Projection: Competition over Control of Newly Opened Sea Routes and Natural Resources

Experts expect the Arctic sea ice extent to continue decreasing. The CMIP5 (Coupled Model Intercomparison Project) model with RCP8.5

---

[25] Pezard et al., 2017, pp. 15–16.

[26] Pezard et al., 2017, pp. 14–15.

[27] Pezard et al., 2017, p. 25.

[28] Andrew Wong, "China: We Are a 'Near-Arctic State' and We Want a 'Polar Silk Road,'" CNBC, February 14, 2018.

[29] Matt McGrath, "China Joins Arctic Council but a Decision on the EU Is Deferred," BBC News, May 15, 2013.

predicts an ice-free September by 2050 with an error estimate of plus or minus ten years;[30] other models predict an ice-free September in the Arctic by 2075 or not at all. Regardless, the shrinking sea ice extent will broaden maritime access and encourage nations to pursue natural resources in the region; it could also spark competition as multiple actors seek the same prizes. Such competition could arise earlier, possibly even by 2030.

Control of Maritime Access

The *Transport Strategy of the Russian Federation Up to 2030* highlights the development of the Northern Sea Route (NSR) as a major Russian national interest.[31] As Figure 3.1 shows, the route lies within Russia's Exclusive Economic Zone (EEZ). In December 2017, the Russian Duma passed a bill, supported by President Vladimir Putin, to grant Russian ships exclusive rights to transport hydrocarbons on the NSR.[32] Not all nations agree that Russia has exclusive jurisdiction over the NSR, and such a move could defy international agreements.[33] Russia is increasing military personnel near the NSR, and will have ten search-and-rescue centers—soon to become military bases—along the route by 2021.[34] Some analysts say these measures indicate that Russia is developing an anti-access presence in the Arctic.[35]

Russia's posture is potentially restrictive in significant ways but does not currently place a stranglehold on global commercial transportation. The NSR offers the fastest possible sea route between many parts of Europe and Asia, but other (albeit longer) paths to conduct

---

[30] T. F. Stocker et al., *Climate Change 2013: The Physical Science Basis*, Geneva: Intergovernmental Panel on Climate Change, 2013, p. 21.

[31] Conley and Rohloff, 2015, p. 83.

[32] Atle Staalesen, "Russian Legislators Ban Foreign Shipments of Oil, Natural Gas and Coal Along Northern Sea Route," *Barents Observer*, December 26, 2017.

[33] Inga Denezh, "Russia Plans to Shut Its Northern Sea Route to Foreign Vessels," *Asia Times*, November 22, 2017.

[34] Conley and Rohloff, 2015, pp. IX, 85; Thomas Nilsen, "Russia Opens New Rescue Base on Northern Sea Route," *Independent Barents Observer*, via January, 12, 2018.

[35] Conley and Rohloff, 2015, p. IX.

Figure 3.1
Arctic Sea Routes and Exclusive Economic Zones

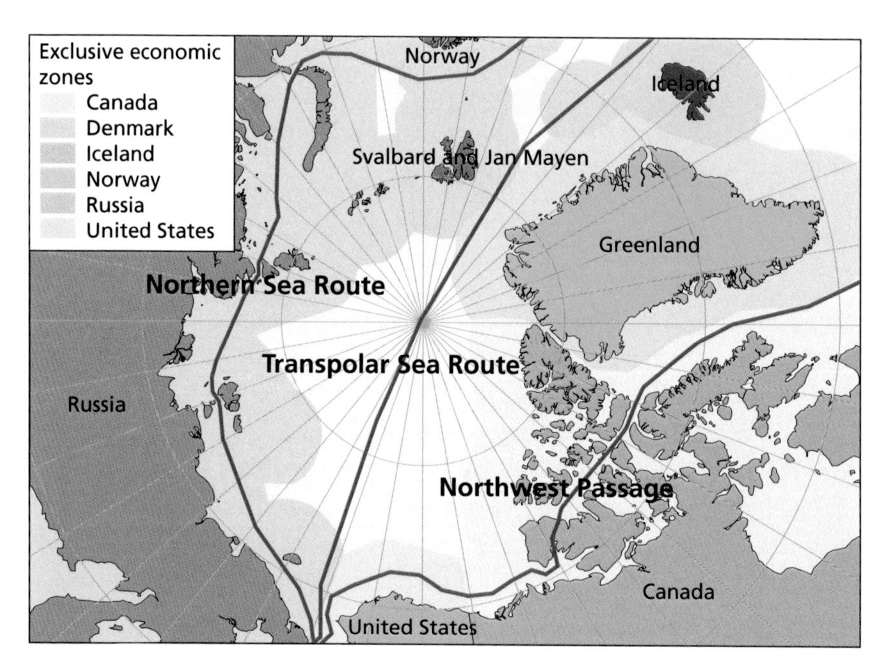

SOURCES: For Arctic sea routes: National Geospatial-Intelligence Agency, "Arctic Sea Routes," January 15, 2016; for countries: Esri, "World Countries," Garmin International, November 30, 2017; for EEZs: Flanders Marine Institute, 2016, Maritime Boundaries Geodatabase, version 10, in conjunction with National Oceanic and Atmospheric Administration, "Exclusive Economic Zones, (EEZs)," May 10, 2018; for graticules: National Oceanic and Atmospheric Administration, National Centers for Environmental Information, "Arctic Graticule," January 18, 2018.

trade do exist, such as the Suez and Panama canals. Furthermore, by 2040, scientists expect that further Arctic ice melt will create summer openings of the even shorter Transpolar Sea Route—a shipping route (Figure 3.1) that crosses over the center of the Arctic Ocean and can stay out of the Russian EEZ completely.[36] In fact, Chinese icebreaker ships began traversing the Transpolar Sea Route in 2012.[37] It follows,

[36] Laurence C. Smith and Scott R. Stephenson, "New Trans-Arctic Shipping Routes Navigable by Midcentury," *Proceedings of the National Academy of Sciences*, Vol. 110, No. 13, 2013.

[37] Laura Zhou, "Slowly but Surely, China Is Carving a Foothold Through the Arctic," *South China Morning Post*, January 26, 2018.

then, that security tensions over controlling the NSR waterway could decrease as the Arctic opens up even further.

Competition for Resources

Previously inaccessible hydrocarbons and minerals are becoming attainable as Arctic ice continues to melt. The Arctic has large quantities of phosphate, bauxite, iron ore, copper, nickel, and fish.[38] It also has vast energy reserves. The U.S. Geological Survey estimates that the Arctic contains 90 billion barrels of oil, 1,669 trillion cubic feet of natural gas, and 44 billion barrels of liquid natural gas.[39] This represents about 22 percent of the world's undiscovered conventional oil and natural gas resource base.[40] Russia requires these new fields to make up for declines in its current production and maintain an output of 10 million barrels of oil per day beyond 2020.[41]

The vast majority of these resources lie in undisputed regions.[42] The disputed areas are less valuable, lowering the potential benefits of winning any conflict over claims. For example, other nations are disputing a claim extended by Russia into the Lomonosov Ridge (Figure 3.2), although the region is expected to contain few hydrocarbons or valuable resources relative to other, undisputed areas of the Arctic.[43] The few resources that might be there likely would be thousands

---

[38] Albert Buixadé Farré et al., "Commercial Arctic Shipping Through the Northeast Passage: Routes, Resources, Governance, Technology, and Infrastructure," *Polar Geography*, Vol. 37, No. 4, 2014, p. 303; Jeremy Mathis, "Fishing in the Arctic?" *NOAA Research News*, September 8, 2017.

[39] Donald L. Gautier and Thomas E. Moore, *Introduction to the 2008 Circum-Arctic Resource Appraisal (Cara) Professional Paper*, Menlo Park, Calif.: U.S. Geological Survey, 2017, p. 8.

[40] Philip Budzik, *Arctic Oil and Natural Gas Potential*, Washington, D.C.: U.S. Energy Information Administration, Office of Integrated Analysis and Forecasting, Oil and Gas Division, 2009, p. 6.

[41] Eurasia Group, *Opportunities and Challenges for Arctic Oil and Gas Development*, Washington, D.C.: Wilson Center, 2013, p. 17.

[42] Brian Beary, "Race for the Arctic," *CQ Global Researcher*, Vol. 2, August 1, 2008.

[43] Gautier and Moore, 2017, p. 8.

Figure 3.2
Overlapping Claims in the Arctic

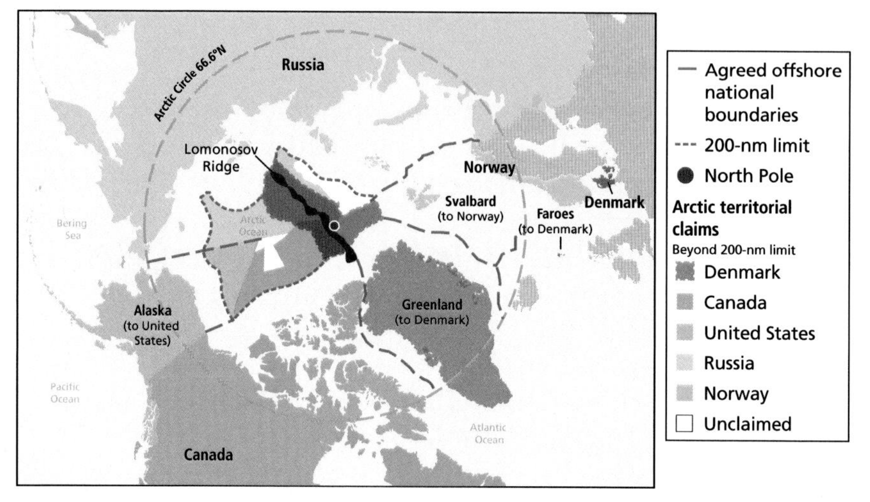

SOURCE: Pezard et al., 2017, p. 42.

of meters below the sea surface and thus not exploitable.[44] Another area of little dispute is commercial fishing: Canada, China, the European Union, Greenland, Iceland, Japan, Norway, Russia, South Korea, and the United States signed an agreement to prohibit this practice in the Arctic until at least 2034.[45]

The effect of the Arctic's extreme environment—shortened workdays, increased labor costs, and other technical difficulties caused by the cold conditions—will also make the newly accessible territory comparatively less profitable than other parts of globe for the foreseeable future. Long lead times (often five to eight years from initiation to full production) make Arctic extraction more complicated and less profitable. For example, an onshore Alaska North Slope project had capital costs that are 50-percent to 100-percent higher than similar oil and

---

[44] Pezard et al., 2017, p. 43.

[45] Timothy Gardner, "Global Powers Strike Deal to Research Before Fishing Arctic Seas," Reuters, November 30, 2017.

natural gas projects in Texas.[46] The Arctic's hydrocarbon base also has much more natural gas than oil; although the former is cheaper to extract, it is also far more expensive to transport over long distances.[47] All in all, the high costs of resource excavation in the Arctic is likely to prevent widespread competition that could lead to fighting over the next few decades.

Implications for the U.S. Air Force and the Future of Warfare

Full-fledged conflict in the Arctic remains unlikely, but economic interests will increase both Russian and Chinese presence there. The USAF will need to monitor and, in some cases, respond to this presence.[48] Russia's aggressive moves suggest that the USAF could be called to the area with increasing frequency to intercept military units that approach U.S. interests in the region and the Alaskan coast.[49]

The Arctic environment also complicates military operations in ways that do not occur in other places on earth. Terrestrial and satellite communications are more limited in this region, particularly above 65° north.[50] The USAF should prepare for an increase in demand for intercept flights and should commit to maintaining or even augmenting its capability to monitor improper foreign activity in the Arctic.[51]

Coastal erosion from melting ice will also threaten U.S. military infrastructure in the Arctic. In some places along Alaska's northern

---

[46] Budzik, 2009, p. 9.

[47] Budzik, 2009, p. 6.

[48] Clay Dillow, "Russia and China Vie to Beat the U.S. in the Trillion-Dollar Race to Control the Arctic," CNBC, February 6, 2018.

[49] Alex Horton, "U.S. Jets Intercept Pair of Russian Bombers off Alaskan Coast," *Washington Post*, May, 12, 2018.

[50] DoD, *Report to Congress on Strategy to Protect United States National Security Interests in the Arctic Region*, Washington, D.C., 2016, p. 13.

[51] Courtney Kube, "Russian Tu-95 Bombers Fly Near Alaskan Coast, Again," NBC News, April 19, 2017.

coast, more than 60 feet of shoreline are lost each year.[52] The USAF has several installations near Alaska's northern coast, including some that are already facing the consequences of coastal erosion. A 2014 report by the Government Accountability Office noted that "the combination of thawing permafrost, decreasing sea ice, and rising sea level on the Alaskan coast have led to an increase in coastal erosion at several USAF radar early warning and communication installations. According to installation officials, this erosion has damaged roads, utility infrastructure, seawalls, and runways."[53] The Cape Lisburne Long Range Radar Station, Oliktok Long Range Radar Site, and Point Barrow Long Range Radar Site will be particularly threatened.[54] These sites are part of the North Warning System, which provides early warning radar for attacks from across the northern polar region.[55] Protecting these facilities will also place an additional budgetary cost on the USAF in years to come. For example, replacing the 5,450-FT rock seawall that protects the installation airstrip at the Cape Lisburne radar site costs $46.8 million, and more expenditures can be expected in coming years.[56]

---

[52] Ann E. Gibbs and Bruce M. Richmond, *National Assessment of Shoreline Change—Summary Statistics for Updated Vector Shorelines and Associated Shoreline Change Data for the North Coast of Alaska, U.S.-Canadian Border to Icy Cape*, Washington, D.C.: U.S. Geological Survey, 2017.

[53] U.S. Government Accountability Office, *DOD Can Improve Infrastructure Planning and Processes to Better Account for Potential Impacts*, Washington, D.C., report to congressional requesters, May 2014, pp. 12–13.

[54] Zachariah Hughes, "Climate Accelerating Erosion at U.S. Radar Facilities in Arctic," *Barents Observer*, July 6, 2016.

[55] Hart Crowser "Environmental Assessment for North Warning System," Washington, D.C.: Department of the Air Force, Air Force Systems Command Electronic Systems Division, November 10, 1986, pp. 2-6, 2-7.

[56] Office of the Under Secretary of Defense for Acquisition, Technology, and Logistics, *Department of Defense Climate-Related Risk to DoD Infrastructure Initial Vulnerability Assessment Survey (SLVAS) Report*, Washington, D.C.: U.S. Department of Defense, January 2018, p. 10.

CHAPTER FOUR

Trend 3: Rising Sea Levels

Rising seas and the increase in frequency and strength of storm surges threaten low-elevation coastal zones. These zones make up only 2 percent of the earth's land area but contain roughly 10 percent of the world's population and 65 percent of cities, with more than 5 million inhabitants.[1] Moreover, these percentages will likely increase over the next two decades.[2] In addition, DoD possesses coastal or tidally influenced sites that could be affected by SLR.[3]

Context: SLR Can Destroy Infrastructure and Livelihoods and Can Prompt Migration

SLR can generate migration from vulnerable areas.[4] This is especially problematic in developing nations, where higher proportions of the population reside in vulnerable low-elevation coastal zones and gov-

---

[1] Anthony Oliver-Smith, "Sea Level Rise and the Vulnerability of Coastal Peoples: Responding to the Local Challenges of Global Climate Change in the 21st Century," UNU-EHS InterSecTions, Vol. 7, 2009, p. 5.

[2] Oliver-Smith, 2009, p. 5.

[3] John A. Hall, Stephen Gill, Jayantha Obeysekera, William Sweet, Kevin Knuuti, and John Marburger, *Regional Sea Level Scenarios for Coastal Risk Management: Managing the Uncertainty of Future Sea Level Change and Extreme Water Levels for Department of Defense Coastal Sites Worldwide*, Alexandria, Va.: Strategic Environmental Research and Development Program, 2016, p. ES-3.

[4] Robert McLeman, "Migration and Displacement Risks Due to Mean Sea-Level Rise," *Bulletin of the Atomic Scientists*, Vol. 74, No. 3, 2018.

ernments frequently lack the capacity and the capabilities to adequately deal with the challenges that SLR presents.[5] SLR also can have a multiplier effect on other migration drivers, such as diminishing economic opportunities, increasing environmental problems, and changing social and political dynamics.[6]

SLR also often acts as a threat multiplier that exacerbates other causes of conflict—such as poverty, political instability, and social tensions.[7] First, migration as a result of SLR can create economic tensions between locals and migrants that could erupt into violent conflict. Second, SLR can expose failures in government and social infrastructure that could undermine governments further and lead to failed states and the emergence of violent nonstate actors. Third, SLR and related flooding can lead to salinization of the water supply that could cause fights over resources.[8]

Historical Trend: Sea Levels Are Rising

Global sea levels have risen 3.2 inches since 1993, double the 20th-century average.[9] SLR, however, varies regionally; the western tropical Pacific, northern Atlantic, and Austral oceans have higher rates of SLR than global averages.[10]

---

5 Oliver-Smith, 2009, p. 24; McLeman, 2018, p. 150.

6 Koko Warner, Charles Ehrhart, Alex de Sherbinin, Susana Adamo, and Tricia Chai-Onn, *In Search of Shelter: Mapping the Effects of Climate Change on Human Migration and Displacement*, Bonn, Germany: United Nations University, CARE, and CIESIN-Columbia University and in close collaboration with the European Commission "Environmental Change and Forced Migration Scenarios Project," the UNHCR, and the World Bank, 2009, p. 2.

7 CNA Military Advisory Board, *National Security and the Accelerating Risks of Climate Change*, Alexandria, Va.: CNA Corporation, 2014, p. 2.

8 CNA Military Advisory Board, *National Security and the Threat of Climate Change*, Alexandria, Va.: CNA Corporation, 2007, pp. 13–28.

9 Rebecca Lindsey, "Climate Change: Global Sea Level," climate.gov, August 1, 2018.

10 Hindumathi Palanisamy, Anny Cazenave, Benoit Meyssignac, Laurent Soudarin, Guy Wöppelmann, and Melanie Becker, "Regional Sea Level Variability, Total Relative Sea Level

The worldwide effects of SLR have been subtle but noticeable. In the past decade, SLR is believed to have contributed to migration with resulting conflict. For example, in the aftermath of the 2010, 2011, and 2012 floods in Pakistan, many Pashtun, Baloch, and rural Sindhi migrated to Karachi. The Muttahida Qaumi Movement (Karachi's dominant political party, made up of the Mohajir minority group) has clashed violently with these new arrivals over land rights and political power.[11] These conflicts, in addition to al Qaeda and the Taliban recruiting and intensifying their attacks in the regions most affected by flooding,[12] threatened to destabilize the government of President Asif Ali Zardari and foment chaos in Pakistan.[13]

Coastal storm surge—which increases in severity with SLR—can lead to flooding, affect infrastructure, and (without proper mitigation measures) make operating in certain locations difficult. Although we are discussing these trends separately, the interaction between SLR and extreme weather events is of particular concern. SLR played a role when Hurricane Isabel flooded Langley Air Force Base in Virginia in 2003 and damaged roughly 22 percent of the base's aircraft engines and produced $166 million in total damages, but it is impossible to determine the exact extent of SLR's contribution to the problem.[14] Similarly, storm surge following in the wake of Hurricane Michael in October 2018 contributed to an unprecedented level of damage to

---

Rise and Its Impacts on Islands and Coastal Zones of Indian Ocean over the Last Sixty Years," *Global and Planetary Change*, Vol. 116, 2014, p. 54.

[11] Tom Wright, "Crush of Refugees Inflames Karachi," *Wall Street Journal*, August 26, 2010.

[12] Ramesh Ghimire and Susana Ferreira, "Floods and Armed Conflict," "Floods and Armed Conflict," *Environment and Development Economics*, Vol. 21, No. 1, 2016, p. 47.

[13] Arpita Bhattacharyya and Michael Werz, *Climate Change, Migration, and Conflict in South Asia*, Washington, D.C.: Center for American Progress, 2012, pp. 20–21.

[14] Emily Atkin, "Inside the U.S. Military's Fight Against Sea Level Rise," *Circa*, January 4, 2017.

Tyndall Air Force Base in Florida, including damage to several (possibly 22) grounded F-22 fighter jets each valued at $100–300 million.[15]

Future Projection: Sea Levels Will Continue to Rise and Lead to More-Frequent Low-Level Conflicts

Using the RCP4.5 emissions scenario and relative to the 1986–2005 levels, global mean sea level is expected to increase between 19 and 33 cm sometime between 2046 and 2065.[16] This could erode coastlines and cause some islands to disappear entirely, which could lead to redrawn maritime boundaries, particularly in the Pacific Ocean and the South China Sea. SLR also could hinder Chinese efforts at building of manmade islands in the region.[17] Already, parts of one such island above the Fiery Cross Reef have fallen into the sea.[18]

SLR will likely contribute in upcoming decades to instability and migration, which will affect U.S. national security interests especially in the Middle East, North Africa, and South Asia.[19] In the Middle East and North Africa (MENA), SLR threatens to erode coastal agricul-

---

[15] John Conger, "Three Takeaways from Hurricane Michael's Impact on Tyndall Air Force Base," *Climate 101 Security*, blog post, October 19, 2018.

[16] Pachauri et al., 2014, p. 60.

[17] *Terriclaiming* refers to the act of territorial reclamation, a term used to describe a nation's reclamation activities as part of a pursuit to preserve or expand territory to augment its geopolitical stature. China's activities in that regard are worrisome to the United States for two reasons: First, China seeks to reclaim territory far away from its national borders, and second, it does so in a clandestine manner. See Wilson T. VornDick, "Thanks Climate Change: Sea-Level Rise Could End South China Sea Spat," *The Diplomat*, November 8, 2012; Wilson T. VornDick, "China's Island Building + Climate Change: Bad News," *Real Clear Defense*, March 9, 2015a; and Wilson T. VornDick, "Terriclaims: The New Geopolitical Reality in the South China Sea," Asia Maritime Transparency Initiative, April 8, 2015b.

[18] Steve Mollman, "It's Typhoon Season in the South China Sea—and China's Fake Islands Could Be Washed Away," *Quartz*, August 1, 2016.

[19] Warner et al., 2009, p. iv; Heather Messera, Ronald Keys, John Castellaw, Robert Parker, Ann C. Phillips, Jonathan White, and Gerald Galloway, *Military Expert Panel Report: Sea Level Rise and the U.S. Military's Mission*, 2nd ed., Washington, D.C.: Center for Climate & Security, 2018, p. 18.

tural areas and salinate freshwater supplies.[20] In South Asia, SLR could cause floods in India, Bangladesh, and Pakistan,[21] and contribute to crop failures, water shortages (because of infiltration of salt water into freshwater aquifers[22]), and displacement and migration in these areas. These effects, combined with existing issues of poor governance, could contribute to internal conflict.

Implications for the U.S. Air Force and the Future of Warfare

SLR can damage USAF infrastructure and impede the ability to conduct operations. A January 2018 DoD survey found that 78 USAF sites (installations and associated structures and facilities) are within 2 kilometers of a coast and 0–6 feet of sea level, including six major installations and numerous mission-critical communications and radar sites.[23] A third of these (26 sites) already have been affected by flooding. All these sites will become more susceptible to SLR in the coming one to two decades.[24]

Another study by the Strategic Environmental Research and Development Program—a joint effort among DoD, the Environmental Protection Agency, the U.S. Department of Energy, and the National Oceanic and Atmospheric Administration—developed a database of regional scenarios for SLR and extreme water levels for DoD sites across the world and determined the sites that were poten-

---

[20] Caitlin E. Werrell and Francesco Femia, eds., *The Arab Spring and Climate Change: A Climate and Security Correlations Series*, Washington, D.C.: Center for American Progress, 2013a, p. 34.

[21] Bhattacharyya and Werz, 2012, pp. 11, 20–21, 24.

[22] Ohio State University, "Climate Change Threatens Drinking Water, As Rising Sea Penetrates Coastal Aquifers," *ScienceDaily*, November 7, 2007.

[23] Office of the Under Secretary of Defense for Acquisition, Technology, and Logistics, 2018.

[24] Office of the Under Secretary of Defense for Acquisition, Technology, and Logistics, 2018, p. 18.

tially deserving of SLR vulnerability or impact assessments by applying a coarse vulnerability screen that classified sites within 20 km of the ocean and below 30 m elevation.[25] Not that all of these sites will necessarily feel the effects of SLR by 2030, but the USAF should consider evaluating the long-term prospects for these sites in the next few decades. As shown in Figure 4.1, the USAF has significant operations at 83 locations within this screen.[26]

SLR threatens Langley Air Force Base, 8 feet above sea level.[27] Langley is home to the Air Combat Command, the majority of F-22s, and the 480th Intelligence, Surveillance, and Reconnaissance (ISR) Wing, which is the core of the USAF's Distributed Common Ground Station and enables the primary ISR planning and execution for global operations.[28] As previously mentioned, the intersection of SLR and extreme weather has already led to flooding that damaged Langley in 2003 during Hurricane Isabel and Tyndall Air Force Base in 2018 during Hurricane Michael.[29] Most of Diego Garcia—an atoll in the Indian Ocean that houses a U.S. bomber base and is an essential hub for operations in Afghanistan, South Asia, and East Africa—is less than 6.5 feet above sea level.[30] The Marshall Islands, where the USAF is building an almost $1 billion radar installation called the "Space Fence" to track artificial satellites and space junk, is only 10 feet above sea level.[31] Eglin Air Force Base in Florida has test facilities and radar

[25] John A. Hall, Stephen Gill, Jayantha Obeysekera, William Sweet, Kevin Knuuti, and John Marburger, *Regional Sea Level Scenarios for Coastal Risk Management: Managing the Uncertainty of Future Sea Level Change and Extreme Water Levels for Department of Defense Coastal Sites Worldwide*, Alexandria, Va.: U.S. Department of Defense, Strategic Environmental Research and Development Program, 2016, pp. 4-5, 4-6.

[26] Hall et al., 2016, pp. 4-5, 4-6.

[27] Messera et al., 2018, p. 38.

[28] Messera et al., 2018, p. 35.

[29] Atkin, 2017; Dave Philipps, "Tyndall Air Force Base a 'Complete Loss' Amid Questions About Stealth Fighters," *New York Times*, October 11, 2018.

[30] Messera et al., 2018, p. 27.

[31] Associated Press, "Rising Seas Could Threaten $1 Billion Air Force Radar Site," CBS News, October 18, 2016.

Figure 4.1
Prominent U.S. Air Force Sites Within 20 km of an Ocean and Below 30 m Elevation

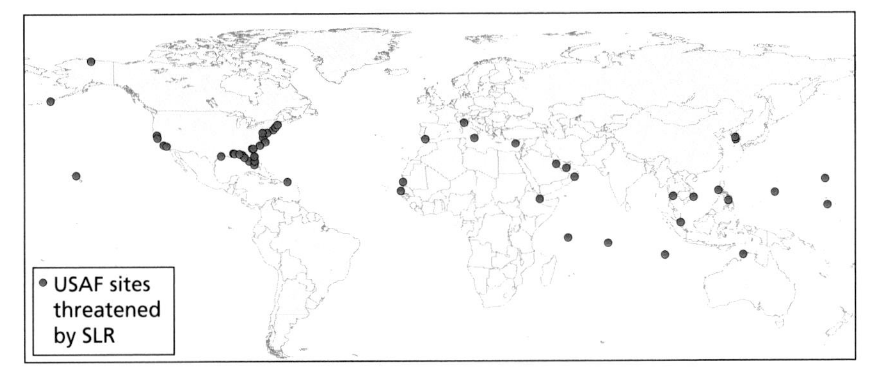

SOURCE: For countries: UNIGIS Geospatial Education Resources, "Countries WGS84," December 6, 2016; for USAF sites in United States and territories: DoD, "Military Installations, Ranges, and Training Areas," data set, January 18, 2017; for USAF sites outside United States: Michael J. Lostumbo, Michael J. McNerney, Eric Peltz, Derek Eaton, David R. Frelinger, Victoria Greenfield, John Halliday, Patrick Mills, Bruce R. Nardulli, Stacie L. Pettyjohn, Jerry M. Sollinger, and Stephen M. Worman, *Overseas Basing of US Military Forces: An Assessment of Relative Costs and Strategic Benefits*, Santa Monica, Calif.: RAND Corporation, RR-201-OSD, 2013; for space surveillance sites: Air Force Space Command, Space Surveillance Network (SSN) Today briefing provided to RAND, undated; for 2012 oceans: Tom Patterson and Nathaniel Vaughn Kelso, vector and raster map data, Natural Earth, undated; for elevation: "USGS National Elevation Dataset (NED)," data.gov, undated; for areas outside United States: National Aeronautics and Space Administration, collections search, undated.

and communications facilities near sea level.[32] Finally, SLR also threatens Andersen Air Force Base in Guam, the only permanent base in the Western Pacific for B-52, B-1, and B-2 heavy strategic bombers, which is several hundred feet above sea level but has its water and energy supplies on low-elevation areas of the island.[33]

The USAF has taken notice. In July 2016, General (ret.) Ron Keys, former commander of Air Combat Command, said, "[W]e have

[32] Messera et al., 2018, p. 28.

[33] Catherine Foley, *Military Basing and Climate Change*, Washington, D.C.: American Security Project, 2012, p. 4; Kyle Mizokami, "Why Is North Korea So Fixated on Guam?" *Popular Mechanics*, August 9, 2017; Mike Ives, "North Korea Aside, Guam Faces Another Threat: Climate Change," *New York Times*, August 11, 2017.

19 bases that we consider jewels in our crown of capability that are going to be affected by sea level rise. And it doesn't have to rise eight feet. It only has to rise a couple of inches, and a good nor'easter pulls in, and all of a sudden we're under water."[34]

SLR also could spark conflict. In the South China Sea, SLR might increase the chances of conflict in the next two decades: China's claims to extend its EEZ depend on low-lying islands that likely will disappear, and seeing this potential loss, China might seek to consolidate these claims quickly, either through legal mechanisms or by force.[35] SLR and the dislocation it causes also could contribute to substate instability, particularly in North Africa, the Middle East, and South Asia. Finally, SLR also could create humanitarian crises, increasing the probability that the USAF will be called to provide HA/DR.[36]

---

[34] Caitlin Werrell and Francesco Femia, "General Keys: The Military Thinks Climate Change Is Serious," Center for Climate and Security, July 7, 2016.

[35] VornDick, 2012.

[36] Messera et al., 2018, p. 18.

CHAPTER FIVE

Trend 4: Extreme Weather Events

Climate change alters the frequency, intensity, spatial extent, duration, and timing of extreme weather events, such as heat waves, tornadoes, hurricanes, tropical cyclones, severe storms (tornadoes, hail, and straight-line winds), wildfires, crop freezes, and winter storms.[1]

Similar to other trends discussed in this report, extreme weather events are driven in part by a steady rise in global temperature. As temperatures continue to rise, more heat and vapor are present in the atmosphere (every 0.5–degree Celsius rise in temperature is associated with an approximately 3-percent increase in atmospheric moisture content). Coupled with higher sea surface temperatures, these conditions accelerate winds and feed storms, thus heightening flood risks.[2] More-extreme and more-frequent weather events interact with and affect other trends discussed in this report, such as SLR, water scarcity, and the rise of megacities. Furthermore, these events could have a significant effect on human populations—and, potentially, on conflict—in years to come.

---

[1] Christopher B. Field, Vicente Barros, Thomas F. Stocker, and Qin Dahe, eds., *Managing the Risks of Extreme Events and Disasters to Advance Climate Change Adaptation: Special Report of the Intergovernmental Panel on Climate Change*, Cambridge, Mass.: New York: Intergovernmental Panel on Climate Change, 2012, p. 111.

[2] National Aeronautics and Space Administration, "The Impact of Climate Change on Natural Disasters," March 3, 2005.

Context: Extreme Weather Events Can Cause Migration, Disease, and Instability

Severe tropical storms and associated flood events generate large volumes of internal displacement and migration. Each of the major storms of the past two decades—e.g., Hurricane Mitch (Central America, 1998), Hurricane Katrina (United States, 2005), Cyclone Aila (Bangladesh, 2009), and Typhoon Haiyan (Philippines, 2013)—was followed by migration. The likelihood of migrants returning home depends on the damage to houses and infrastructure and the extent of recovery.[3] Beyond economic dimensions, forced large-scale migration is a source of friction between locals and newcomers.[4]

Extreme weather events also can increase the risk and spread of disease. The Gulf Coast of the southern United States—home to several critical military bases and sites—is increasingly at risk of infectious diseases, partly because "periodic exposures to climate and environmental hazards, including hurricanes, floods, droughts, and oil spills," amplify such risk factors as poverty and poor sanitation.[5] Extreme weather can also increase the spread of such diseases as malaria, Dengue fever, and Zika that are transmitted by mosquitoes, ticks, fleas, and sandflies, and that account for more than 17 percent of all infectious diseases and cause more than a million deaths annually.[6] For example, increased precipitation expands the range of insect breeding sites, bringing more mosquitoes near more people.[7] The Zika epidemic of 2015–2016 is an

---

[3] Caitlin E. Werrell and Francesco Femia, eds., *Epicenters of Climate and Security: The New Geostrategic Landscape of the Anthropocene*, Washington, D.C.: Center for Climate and Security, June 2017.

[4] World Bank, "High and Dry: Climate Change, Water, and the Economy," Washington, D.C., 2016.

[5] Peter Hotez, Kristy O. Murray, and Pierre Buekens, "The Gulf Coast: A New American Underbelly of Tropical Diseases and Poverty," *PLoS Neglected Tropical Diseases*, Vol. 8, No. 5, 2014.

[6] World Health Organization, "Vector-Borne Diseases," October 31, 2017.

[7] Andrew Githeko, Steve W. Lindsay, Ulisses E. Confalonieri, and Jonathan A. Patz, "Climate Change and Vector-Borne Diseases: A Regional Analysis," *Bulletin of the World Health Organization*, Vol. 78, No. 9, 2000.

illustrative example. Models found that climate conditions, including an unusually mild 2015 El Niño season, were optimal for a large-scale mosquito-borne transmission of Zika.[8]

Finally, as will be discussed in the next chapter, extreme weather events—primarily droughts and floods—are often followed by spikes in violence, civil war, and regime change in developing countries, leading to increased economic grievances, decreased government capacity, or some combination thereof.[9]

Historical Trend: The Risks from Extreme Weather Events Have Become More Severe

Globally, the number of floods and other hydrological events have quadrupled since 1980 and have doubled since 2004 (Figure 5.1).[10] Between 1960 and 1990, floods in Europe destroyed assets worth $7 billion per year, on average.[11] A study that examined tropical cyclones during the years 1950–2008 found that national incomes do not recover for 20 years after such a disaster.[12] The United States has experienced more than 200 such extreme events since 1980 at a total cost exceed-

---

[8] Cyril Caminade, Joanne Turner, Soeren Metelmann, Jenny C. Hesson, Marcus S. C. Blagrove, Tom Solomon, Andrew P. Morse, and Matthew Baylis, "Global Risk Model for Vector-Borne Transmission of Zika Virus Reveals the Role of El Niño 2015," *Proceedings of the National Academy of Sciences*, Vol. 114, No. 1, 2016.

[9] Markus Brückner and Antonio Ciccone, "Rain and the Democratic Window of Opportunity," *Econometrica*, Vol. 79, No. 3, 2011.

[10] European Academies' Science Advisory Council, Leopoldina—Nationale Akademie der Wissenschaften, "New Data Confirm Increased Frequency of Extreme Weather Events: European National Science Academies Urge Further Action on Climate Change Adaptation," *ScienceDaily*, March 21, 2018.

[11] Stephane Hallegatte, Mook Bangalore, Laura Bonzanigo, Marianne Fay, Tamaro Kane, Ulf Narloch, Julie Rozenberg, David Treguer, and Adrien Vogt-Schilb, *Shock Waves: Managing the Impacts of Climate Change on Poverty*, Washington, D.C.: World Bank, 2015.

[12] Solomon M. Hsiang and Amir S. Jina, *The Causal Effect of Environmental Catastrophe on Long-Run Economic Growth: Evidence from 6,700 Cyclones*, Cambridge, Mass.: National Bureau of Economic Research, NBER Working Paper, No. w20352, 2014.

ing $1.1 trillion.[13] In the fall of 2013, a wave of "super-typhoons" with winds greater than 150 mph were recorded in the Central Pacific. Among these was Super-Typhoon Haiyan, the most extreme storm ever recorded to make landfall and the deadliest typhoon in Philippine history, causing more than 7,000 fatalities.[14] The latter prompted a large-scale U.S. military HA/DR mission, involving more than 13,000 soldiers, sailors, airmen, and Marines.[15]

Recent history also illustrates the effects of extreme weather events on migration. According to the International Disaster Monitoring Centre, 26 million people per year have been displaced by disasters on average since 2008.[16] Furthermore, detailed regression analysis from 2011 has shown that extreme weather events and changing climate conditions lead to higher migration from Algeria, Egypt, Morocco, Syria, and the Republic of Yemen.[17] Most international migration occurs at the subnational and subregional levels, primarily among less-developed countries.[18]

---

[13] National Centers for Environmental Information, "Billion-Dollar Weather and Climate Disasters," undated.

[14] Andrew Holland, "How Focusing on Climate Security in the Pacific Can Strengthen Alliances: Lessons from the Global Defense Index on Climate Change for the U.S.," in Caitlin E. Werrell and Francesco Femia, eds., *The U.S.–Asia Rebalance, National Security and Climate Change*, Washington, D.C.: Center for American Progress, November 2015; Jeff Masters, "Super Typhoon Haiyan Storm Surge Survey Finds High Water Marks 46 Feet High," *Weather Underground*, May 8, 2014.

[15] Cheryl Pellerin, "Recovery Effort Takes on Great Energy, Task Force Commander Says," DoD News, November 19, 2013.

[16] International Displacement Migration Centre, *Global Estimates 2015: People Displaced by Disasters*, Geneva, 2015.

[17] Quentin Wodon, Andrea Liverani, George Joseph, and Nathalie Bougnoux, *Climate Change and Migration: Evidence from the Middle East and North Africa*, Washington, D.C.: World Bank, 2014, p. 15.

[18] Stephanie J. Nawyn, "Migration in the Global South: Exploring New Theoretical Territory," *International Journal of Sociology*, Vol. 46, No. 2, 2016.

Figure 5.1
Trends in Different Natural Disaster and Extreme Weather Events

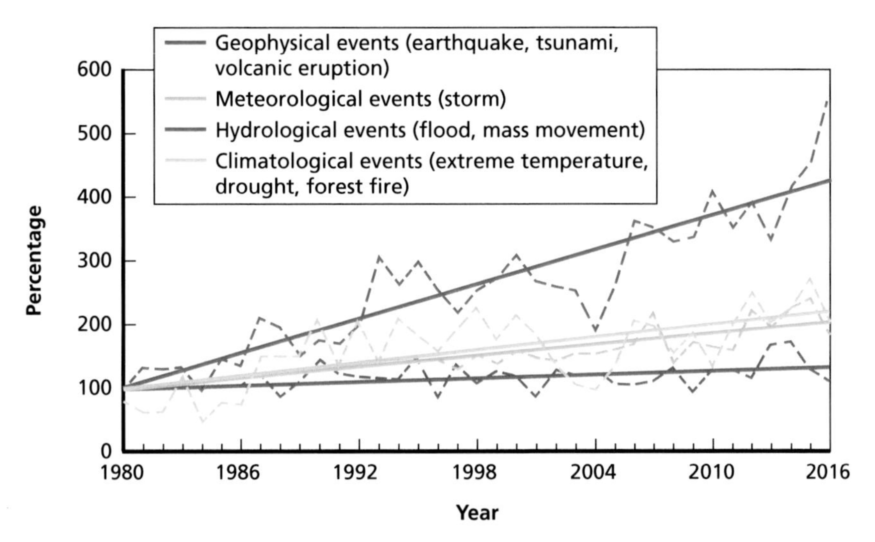

SOURCE: European Academies' Science Advisory Council, Leopoldina—Nationale Akademie der Wissenschaften, 2018. Used with permission.

Future Projection: Extreme Weather Events Will Create Risks for Coastal and Riverine Communities and Infrastructure, Including Military Installations

Extreme weather events are projected to increase in severity and frequency over the next several decades.[19] As the amount of heat in the atmosphere increases, the frequency and intensity of extreme precipitation events will increase in many places. Places that are already wet will become wetter, dry places will generally become drier, making floods and droughts more common.[20] Increasing storm surges on already rising ocean levels will leave coastal communities at risk. Flood hazards are projected to increase in more than half of the world—not only on coastal sea areas but also around river basins. Models present the same

---

[19] Field et al., 2012; Pachauri et al., 2014, pp. 1–32.

[20] Great Lakes Integrated Sciences + Assessments, "Extreme Precipitation," webpage, undated.

broad global and regional trends, but they do not provide reliable local-scale projections of future precipitation patterns.[21] Still, some models predict increasing flood hazards in parts of South Asia, Southeast Asia, East Africa, Central and West Africa, Northeast Eurasia, and South America.[22]

Increasingly severe tropical storms and associated flood events are likely to generate larger volumes of internal displacement and migration.[23] The greatest concentrations of displaced people in the future will likely be in (1) dryland regions with highly seasonal precipitation fluctuations, (2) heavily populated low-lying coastal areas that regularly experience tropical cyclone activity, and (3) atolls in the Pacific and Indian oceans (Figure 5.2).[24]

The increased risks from extreme weather in the future could jeopardize both coastal infrastructure (such as railroads, ports, airports, roads, power and water supplies, and storm and sewage water management) and maritime straits (such as the Panama Canal). In June 2016, the Panama Canal was deepened and widened to enable larger vessels to use it, but it is sensitive to water levels in Panama's Gatun Lake, which are affected by El Niño events.[25] Climate effects could lead to

---

[21] Yasuaki Hijioka, Erda Lin, and Joy Jacqueline Pereira, "Asia," in Vicente R. Barros and Christopher B. Field, eds., *Climate Change 2014: Impacts, Adaptation, and Vulnerability,* Part B: *Regional Aspects; Working Group II Contribution to the Fifth Assessment Report of the Intergovernmental Panel on Climate Change*, New York: Cambridge University Press, 2014.

[22] Blanca E. Jiménez Cisneros, Taikan Oki, Nigel W. Arnell, Gerardo Benito, J. Graham Cogley, Petra Doll, Tong Jiang, and Shadrack S. Mwakalila, "Freshwater Resources," in Christopher B. Field and Vicente Barros, eds., *Climate Change 2014: Impacts, Adaptation, and Vulnerability,* Part A: *Global and Sectoral Aspects; Working Group II Contribution to the Fifth Assessment Report of the Intergovernmental Panel on Climate Change*, New York: Cambridge University Press, 2014.

[23] Kevin J. E. Walsh, John L. McBride, Philip J. Klotzbach, Sethurathinam Balachandran, Suzana J. Camargo, Greg Holland, Thomas R. Knutson, James P. Kossin, Tsz-cheung Lee, Adam Sobel, and Masato Sugi, "Tropical Cyclones and Climate Change," *WIREs Climate Change*, Vol. 7, No. 1, January–February 2016.

[24] Robert McLeman, "Migration and Displacement in a Changing Climate," in Werrell and Femia, 2017.

[25] United Nations Office for Disaster Risk Reduction, "Easing Impact of Drought on the Panama Canal," June 28, 2016.

Figure 5.2
Extreme Weather (Floods, Cyclones, and Mudslides) Hazard Risk

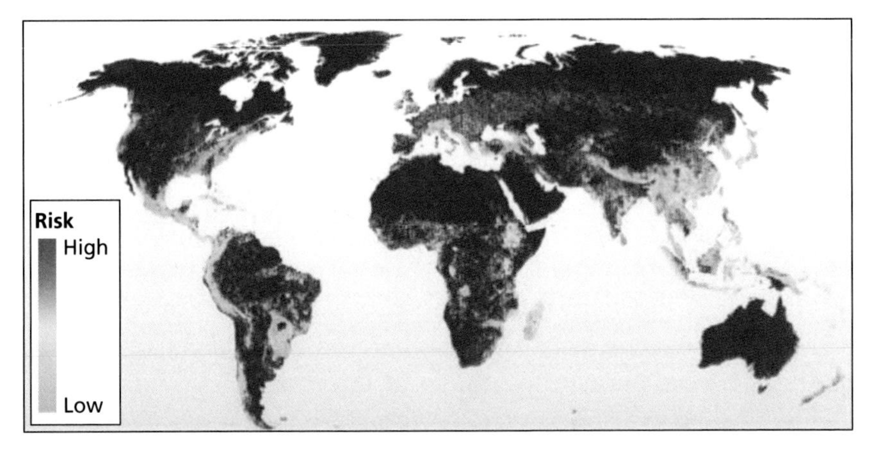

SOURCE: World Water Assessment Program, *Managing Water Under Uncertainty and Risk*, Paris: United Nations Educational, Science, and Cultural Organization, World Water Development Report 4, 2012. Used with permission.

increased rainfall and runoff in the Panama Canal region from May to December, affecting operations.[26]

Implications for the U.S. Air Force and the Future of Warfare

More-frequent and more-severe extreme weather events will place a greater burden on military units, personnel, and assets located in disaster areas or tasked with responding to such events in the United States and abroad.[27] Heavier and more-frequent precipitation will also affect installation maintenance costs and require additional flood or erosion

---

[26] Panama National Environmental Authority, *Enhancing Resilience to Climate Change and Climate Variability in the Central Pacific Region of Panama*, April 2013.

[27] Office of the Under Secretary of Defense for Acquisition, Technology, and Logistics, 2018, p. 7.

control measures.[28] As discussed, Langley and Tyndall have already experienced flooding as a result of a combination of factors, including rising sea levels. In addition, Joint Base Anacostia-Bolling, located at the juncture of the Potomac and Anacostia rivers in Southeast Washington, D.C., hosts the Defense Intelligence Agency headquarters.[29] Low-lying areas of the joint base are already affected by flooding 43 times per year on average. By 2050, such flooding could occur twice per day, depending on tidal flooding scenario. If adaptations are not made well in advance, communication infrastructure and military housing could be at risk.[30] Bases in the Asia-Pacific region will be similarly affected by this trend for two reasons: They are at risk of extreme weather events themselves, and they also would likely serve as hubs for HA/DR missions in particularly hard-hit areas of Asia.[31] Finally, the link between extreme weather and infectious disease means the USAF will need to be prepared to protect service members' health, particularly as they are called to respond to disaster-stricken areas.

---

[28] DoD, *FY 2012 Climate Change Adaptation Roadmap*, Washington, D.C., September 18, 2012.

[29] Commander, Naval Installations Command, "Joint Base Anacostia-Bolling—About," webpage, undated; Defense Intelligence Agency, "About DIA," webpage, undated.

[30] Erika Spanger-Siegfried, Kristina Dahl, Astrid Caldas, and Shana Udvardy, "The U.S. Military on the Front Lines of Rising Seas: Exposure to Coastal Flooding at Joint Base Anacostia-Bolling and Washington Navy Yard, Washington, District of Columbia," Cambridge, Mass.: Union of Concerned Scientists, fact sheet, July 2016.

[31] Constantine Samaras, "U.S. Military Basing Considerations During a Rebalance to Asia: Maintaining Capabilities Under Climate Change Impact," in Caitlin E. Werrell and Francesco Femia, eds., *The U.S.–Asia Rebalance, National Security and Climate Change*, Washington, D.C.: Center for American Progress, November 2015,.

CHAPTER SIX

Trend 5: Growing Water Scarcity

Water is considered "the pillar on which global security, prosperity, and equity stand."[1] Currently, 3.6 billion people—almost half of the world's population—live in areas with potential water stress for one month a year.[2] By 2050, this number could rise to 4.8–5.7 billion people if current trends continue and better water-management practices are not adopted.[3] About 70 percent of people affected live in Asia.[4] Indeed, from 2012 to 2018, the World Economic Forum's Global Risks Perception Survey identified growing water scarcity as one of the top five risks to global stability.[5] As with extreme weather events, however, water scarcity is considered only one factor of many that can lead to

---

[1] William Hague, "The Diplomacy of Climate Change," Russell C. Leffingwell Lecture, Council on Foreign Relations, New York, September 27, 2010.

[2] Mesfin M. Mekonnen and Arjen Y. Hoekstra, "Four Billion People Facing Severe Water Scarcity," *Science Advances*, Vol. 2, No. 2, 2016.

[3] World Water Assessment Program, *Nature-Based Solutions for Water*, Paris: United Nations Educational, Science, and Cultural Organization, World Water Development Report, 2018.

[4] Peter Burek, Yusuke Satoh, Günther Fischer, Mohammed Taher Kahil, Angelika Scherzer, Syliai Tramberend, Luzma Fabiola Nava, Yoshihide Wada, Stephanie Eisner, Martina Florke, Naota Hanasaki, Piotr Magnuszewski, Bill Cosgrove, and David Wiberg, *Water Futures and Solution: Fast Track Initiative*, Laxenburg, Austria: International Institute for Applied Systems Analysis, Working Paper WP-16-006, 2016.

[5] This is an annual survey in which some 1,000 experts and decisionmakers assess the likelihood and impact of 30 global risks over a ten-year horizon, World Economic Forum, *Global Risks Report 2018*, 13th ed., Geneva, 2018.

instability, often when social and political tensions already exist and institutional capacity is limited.

Context: Water Scarcity Could Lead to Intrastate (Not Interstate) Wars

Records point to some 4,000 water conflicts dating back to 3000 BCE.[6] Although several regions remain prone to interstate water conflict, some scholars point out that transboundary water relationships are governed by compromise and cooperation more often than they are by war.[7] In the post–Cold War era, interstate water wars became rare. In the early 2000s, the weight of scholarly opinion shifted away from treating water per se as a *casus belli* for interstate wars and toward examining how water shortages could fuel intrastate violence and state failure, especially when combined with substantial political, ethnic, or religious tensions and the absence of institutions that can promote cooperation.[8] In 2012, the National Intelligence Council concluded that water insecurity could lead to instability and state failure and exacerbate subnational and regional tensions.[9] Research has shown that droughts are often followed by spikes in violence and civil war. In sub-Saharan Africa, for example, evidence indicates that civil wars are more likely to erupt after periods of low rainfall.[10] In rural areas of Brazil, violent land invasions and conflict are more

---

[6] Pacific Institute, "Water Conflict Chronology List," webpage, 2018.

[7] Mathias Albert, Stephan Stetter, Eva Herschinger, and Thomas Teichler, "Conflicts About Water: Securitizations in a Global Context," *Cooperation and Conflict*, Vol. 46, No. 4, 2011, p. 444; Patrice C. McMahon, "Cooperation Rules: Insights on Water and Conflict from International Relations," in Jean Axelron Cahan, ed., *Water Security in the Middle East: Essays in Scientific and Social Cooperation*, London: Anthem Press, 2017, p. 28.

[8] Albert et al., 2011, p. 444.

[9] U.S. Department of State, "Global Water Security 2012," Intelligence Community Assessment ICA 2012-08, February 2, 2012.

[10] Edward Miguel, Shanker Satyanath, and Ernest Sergenti, "Economic Shocks and Civil Conflict: An Instrumental Variables Approach," *Journal of Political Economy*, Vol. 112, No. 4, 2004.

common during dry years.[11] Similarly in India, violence over property is more likely in years with low rainfall.[12]

In regions where agriculture is the main source of employment, water scarcity can lead to poverty, weak governments, and political instability.[13] In addition, water shortages could create upward pressure on water and food prices. Food insecurity, which can result directly from water scarcity, has also been shown to affect individual decision-making regarding participation in violence. Nonstate actors proliferate and become more influential when states are weak and unable to provide food and water.[14] In such environments, nonstate actors, including terrorist organizations, seize water and food resources and weaponize them in the form of leverage that can further erode the legitimacy of the state.[15] One example is the insurgent groups in northern Mali who exploited food insecurity to enlist locals by promising them "food for Jihad."[16] Migration within and between countries, which tends to increase in areas facing acute water shortages, is a widely documented source of friction between groups.[17]

---

[11] Hidalgo et al., 2010.

[12] Heather Sarsons, "Rainfall and Conflict: A Cautionary Tale," *Journal of Development Economics,* Vol. 115, No. C, 2015.

[13] Brückner and Ciccone, 2011.

[14] Todd Sandler, "The Analytical Study of Terrorism Taking Stock," *Journal of Peace Research*, Vol. 51, No. 2, 2014.

[15] Marcus DeBois King, "The Weaponization of Water in Syria and Iraq," *Washington Quarterly*, Vol. 38, No. 4, Winter 2016.

[16] Chris Arsenault, "Drought, Expanding Deserts and 'Food for Jihad' Drive Mali's Conflict," Reuters, April 27, 2015.

[17] Hsiang, Burke, and Miguel, 2013; Andrew R. Solow, "Global Warming: A Call for Peace on Climate and Conflict," *Nature*, Vol. 497, 2013.

Historical Trend: Water-Related Conflict Has Been on the Rise

Despite advancements in technology, water supply has remained limited and demand for water has increased substantially in the developing world (Figure 6.1), primarily because of population growth, more-intense food production, changing consumption habits, and urbanization.[18] The world's population tripled in the 20th century, but the use of water increased sixfold, with irrigated agriculture accounting for 70 percent of water withdrawals.[19]

Increased demand without commensurate changes in water management or increases in supply has led to water shortages that affect half of the world's population.[20] Half a billion people face severe water scarcity—defined by the Global Water Forum as the lack of access to sufficient water for humans and the environment[21]—all year round: 180 million in India, 73 million in Pakistan, 27 million in Egypt, 20 million in Mexico, 20 million in Saudi Arabia, and 18 million in Yemen.[22] Most of the populations of Libya, Somalia, Pakistan, Morocco, Niger, and Jordan also experience year-round severe water scarcity.[23]

---

[18] Only 3 percent of global water is considered fresh water, or salt-free. Nearly 70 percent of fresh water is locked in glaciers and icebergs, thus unavailable. Freshwater supply comes from rain or from rivers, lakes, and springs. In addition, groundwater is available from aquifers, most of which are recharged by the annual water cycle. But depletion rates outpace rates of recharge in many regions, hurting both water quantity and quality (because of infiltration of seawater and wastewater). Water treatment technologies could increase water supply but are costly and energy-intensive, and they are not always viable—e.g., landlocked countries are limited in desalination potential. See World Bank, 2016.

[19] World Bank, 2016. It is important to note that in principle, even small improvements in irrigation efficiency could have meaningful effects.

[20] World Water Assessment Program, 2018.

[21] Chris White, "Understanding Water Scarcity: Definitions and Measurements," Global Water Forum, May 7, 2012.

[22] Mekonnen and Hoekstra, 2016, p. 3.

[23] Mekonnen and Hoekstra, 2016, p. 3.

Figure 6.1
Global Water Withdrawal by Sector, 1900–2010

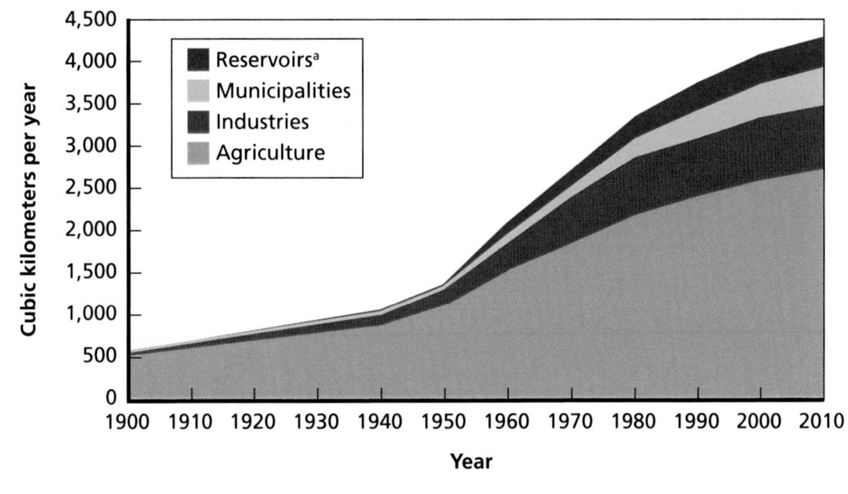

SOURCE: Food and Agriculture Organization of the United Nations, "Maps and Spatial Data," webpage, undated. Used with permission.
[a] Evaporation from artificial lakes.

The mismatch between water supply and demand can be addressed by nations' adaptive capacity—their ability to accommodate long-term changes and adapt to short-term water disruptions.[24] Thus, many countries—indeed, most people in the world—have suffered water shortages in recent years. The consequences of this trend in the short term primarily concern countries that lack adaptive capacity, a nation's ability to provide resources over time and in response to disruptions.[25] As the map in Figure 6.2 indicates, these are mainly the

---

[24] Henry H. Willis, David G. Groves, Jeanne S. Ringel, Zhimin Mao, Shira Efron, and Michele Abbott, *Developing the Pardee RAND Food-Energy-Water Security Index*, Santa Monica, Calif.: RAND Corporation, TL-165-RC, 2016.

[25] Willis et al., 2016. Adaptive capacity reflects the potential for developing new sources of water using domestically available sustainable resources. The index described by Willis et al. measured adaptive capacity by evaluating the total per capita internally available renewable water, which is the sum of internal renewable water resources and external actual renewable water resources.

Figure 6.2
Water Adaptive Capacity by Country

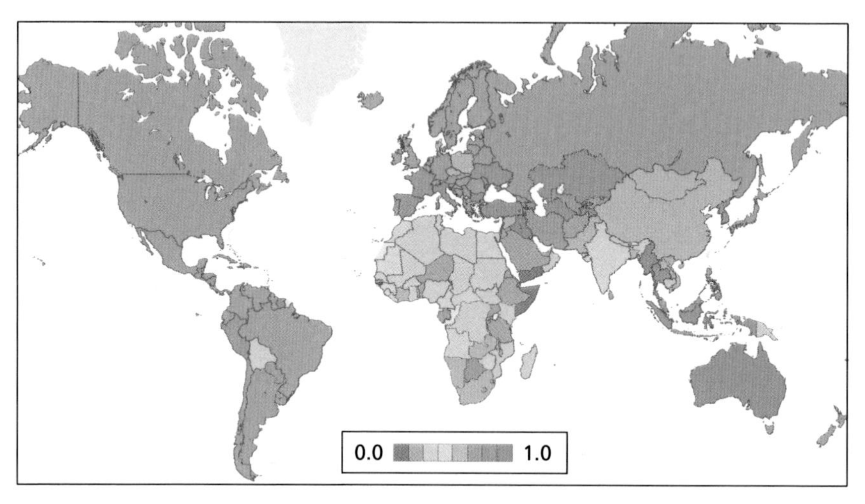

SOURCE: Willis et al., 2016.

countries of the MENA region, sub-Saharan Africa, and Central and South Asia.

The frequency of conflicts in which access to water or water scarcity has been cited as a contributor to violence or where participants have used control over water as a tactic in war has risen in the past three decades (Figure 6.3).[26] In the Middle East and Africa, these conflicts have grown in frequency and intensity. More than 40 such conflicts occurred in the region from 2011, early in the Arab Spring, to 2017: Of these, 12 were in Syria, seven in Yemen, and six in Iraq. In Africa, 37 water conflicts were recorded during the same period, with six in Kenya, four in Somalia, and four in South Sudan and Darfur.[27]

Water shortages indirectly might have helped trigger the Syrian civil war. War erupted because of a complicated array of intertwined

---

[26] Peter Gleick, "Water, Security, and Conflict: Violence over Water in 2015," *ScienceBlogs*, February 17, 2016.

[27] Pacific Institute, 2018.

Figure 6.3
Water-Related Conflict Events: 1930–2015

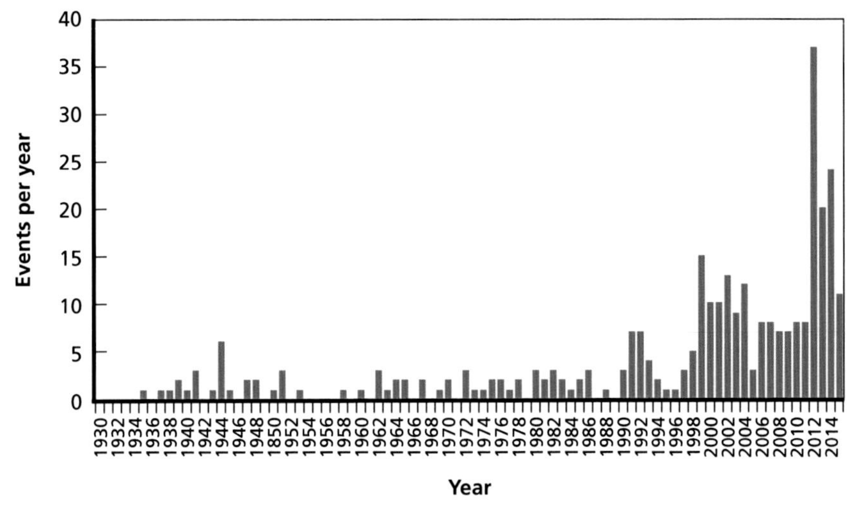

SOURCE: Gleick, 2016. Used with permission.

social, political, economic, and regional factors.[28] But a severe drought from 2006 to 2011 combined with unsustainable water practices caused agricultural failures that led Syria's economy to deteriorate and 1.5 mil-lion Syrians to move from rural areas into such cities as Aleppo, Deir ez-Zour, Hama, Homs, and Dara'a.[29] This migration strained socio-economic and political infrastructure and created the underlying con-ditions that allowed Syria to collapse into civil war.[30]

---

[28] Christiane J. Fröhlich, Jan Selby, Omar S. Dahi, and Mike Hulme, "Climate Change and the Syrian Civil War Revisited," *Political Geography*, Vol. 60, 2017.

[29] Caitlin Werrell and Francesco Femia, eds., *Climate Change Before and After the Arab Awakening: The Cases of Syria and Libya*, Washington, D.C.: Center for Climate and Security, February 2013b; Peter H. Gleick, "Water, Drought, Climate Change, and Conflict in Syria," *Weather, Climate, and Society*, Vol. 6, No. 3, 2014; Colin P. Kelley, Shahrzad Mohtadi, Mark A. Cane, Richard Seager, and Yochanan Kushnir, "Climate Change in the Fertile Crescent and Implications of the Recent Syrian Drought," *Proceedings of the National Academy of Sciences*, Vol. 112, No. 11, 2015.

[30] World Food Program, *Syria Arab Republic Joint Rapid Food Security Needs Assessment (JRFSNA)*, Rome: United Nations Food and Agriculture Organization, 2012.

Parallel events have occurred in sub-Saharan Africa. In the Middle Belt of Nigeria, the semi-nomadic Muslim Fulani tribe have clashed with predominantly Christian farmers as water shortages have increasingly limited grazing lands.[31] The 2015 Global Terrorism Index identified the "Fulani militants" as one of the deadliest terrorist groups in the world.[32] In Somalia, a 2011 drought and Al Shabab's decision to cut off the water supply to cities liberated by the Somali government led to more than 250,000 deaths and hundreds of thousands of displaced persons.[33]

Future Projection: Water Shortages Will Exacerbate Tensions in Vulnerable Regions

Climate change is likely to make global water supply more variable and unpredictable in years to come.[34] Demand for water will increase substantially as global population grows and a new middle class demands more water-intensive foods and more energy;[35] these trends could lead to half of the global population suffering extreme water shortages by 2025.[36] Without intervention, municipal and industrial water demand could increase by around 50 percent by 2030; water demand for the global food system could grow by as much as 40 to 50 percent;[37] and

---

[31] Peter Akpodiogaga and Odjugo Ovuyovwiroye, "General Overview of Climate Change Impacts in Nigeria," *Journal of Human Ecology*, Vol. 29, No. 1, 2010.

[32] Rose Troup Buchanan, "Global Terrorism Index: Nigerian Fulani Militants Named as Fourth Deadliest Terror Group in World," *Independent*, November 18, 2015.

[33] International Organization of Migration, Department of Operations and Emergencies, *Dimensions of Crisis on Migration in Somalia*, Geneva, working paper, February 2014.

[34] Jiménez Cisneros et al., 2014.

[35] Golam Rasul, "Managing the Food, Water, and Energy Nexus for Achieving the Sustainable Development Goals in South Asia," *Environmental Development*, Vol. 18, 2016.

[36] World Health Organization, "Drinking-Water," fact sheet, February 7, 2018.

[37] 2030 Water Resources Group, *Charting Our Water Future: Economic Frameworks to Inform Decision-Making*, New York, 2009.

energy sector water demand could increase by 85 percent.[38] Unless efficiency improves substantially, global demand for water could exceed supply by 40 percent by 2030.[39] If this occurs, regions already experiencing water shortages—including many countries in the MENA region, East and South Africa, Central Asia, and Central America—will suffer even more dire scarcities.[40] Figure 6.4 illustrates projected changes in water scarcity from 2010 to 2050 assuming that social, economic, and technological conditions continue on their current trajectories.

As Figure 6.4 shows, much of the change in water stress is projected in developing countries, including practically the whole MENA region, Central Asia, and parts of South Asia, which are already suf-

Figure 6.4
Water Stress by Country in 2040 Under Business-as-Usual Scenario

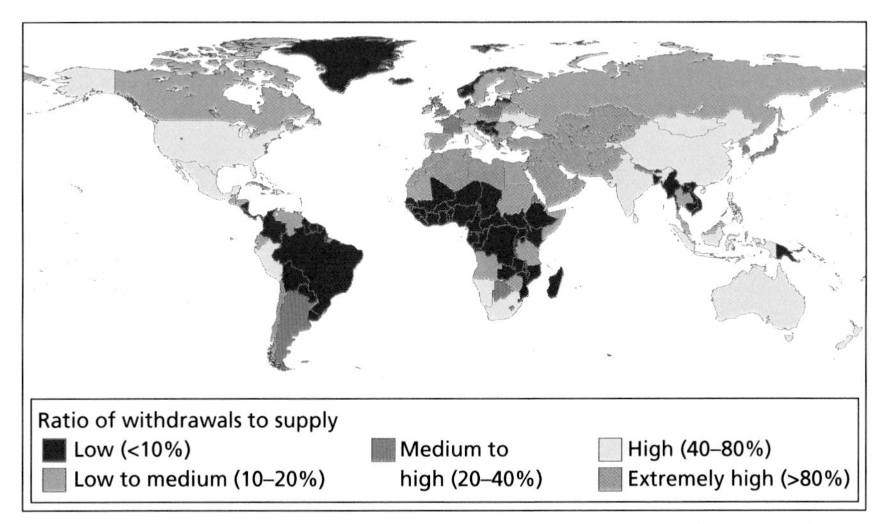

SOURCES: For countries: Esri, 2017; for water scarcity: Tianyi Luo, Robert Samuel Young, and Paul Reig, "Aqueduct Projected Water Stress Country Rankings," Washington, D.C., World Resources Institute, data set, August 2015.
NOTE: This scenario is based on the Shared Socioeconomic Pathway (SSP2) and RCP8.5.

---

[38] World Bank, 2016.

[39] 2030 Water Resources Group, 2009.

[40] World Bank, 2016.

fering from chronic water shortages. The vulnerability to acute water shortages will be especially pronounced in areas where subsistence agriculture is the main source of employment. Currently, 800 million people, 78 percent of the world's population, rely on agriculture for their livelihood,[41] and rainfall changes could reduce crop yield globally by 10 percent by 2030.[42]

The Middle East is particularly vulnerable. Fourteen of the 33 countries at risk of the greatest water stress in 2040 are in the MENA region. Most countries in that region are dependent on food imports and thus particularly exposed to price shocks. In 2007–2008, a prolonged drought affecting major grain-exporting countries—including Ukraine, Europe, and Australia—led to a substantial hike in the price of rice and wheat. Bread subsidies in Egypt were insufficient to prevent price increases in rural areas and led to unrest.[43] Import dependence in the MENA region is projected to rise to 63 percent in 2030 (from 56 percent in 2000), increasing regional vulnerability to high and unpredictable global food prices.[44] Given that the Middle East and North Africa are already prone to terrorism and political instability, the compounding stress of water shortages could inflame an already explosive situation. Unsurprisingly, Middle East security experts often rank water scarcity as one of the main factors that could lead to instability in the region.[45]

---

[41] Petr Havlík, Hugo Valin, Mykola Gusti, Erwin Schmid, Nicklas Forsell, Mario Herrero, Nikolay Khabarov, Aline Mosnier, Matthew Cantele, and Michael Obersteiner, *Climate Change Impacts and Mitigation in the Developing World: An Integrated Assessment of the Agriculture and Forestry Sectors*, Washington, D.C.: World Bank Group, Policy Research Working Paper No. WPS 7477, 2015.

[42] Havlík et al., 2015.

[43] Brian Wright and Carlo Cafiero, "Grain Reserves and Food Security in the Middle East and North Africa," *Food Security*, Vol. 3, Suppl. 1, February 2011, p. 61; Troy Sternberg, "Chinese Drought, Bread and the Arab Spring," *Applied Geography*, Vol. 34, No. 4, May 2012.

[44] Wright and Cafiero, 2011.

[45] Authors' interviews with senior Jordanian military officers, Jordanian scholars, and official U.S. personnel, Amman, Jordan, May 12–13, 2018.

Implications for the U.S. Air Force and the Future of Warfare

Growing water scarcity could have both direct and indirect implications for the future of warfare and the USAF. Growing water insecurity in areas that are already vulnerable might prompt more U.S. military interventions, both HA/DR missions and otherwise. This has happened before: Operation Restore Hope in Somalia in 1992 is one example.[46] Similarly, bases in regions with water scarcity would have to cope with this limitation. The USAF will need to adopt more-efficient uses of energy and water to reduce its own dependence on unreliable supplies of water abroad, especially in regions afflicted with acute water shortages, such as the Middle East.

---

[46] Travis Tritten, "When Disaster Strikes, U.S. Military Assets Often Key to Relief Efforts," *Stars and Stripes*, November 16, 2013.

CHAPTER SEVEN

Trend 6: Increasing Urbanization and Megacities

The world is becoming increasingly urbanized. For the first time in 2008, more than half of the world's population lived in cities.[1] Not only that, they are living in very big cities. *Megacities*, urban areas with more than 10 million people, generate about 14 percent of global gross domestic product.[2] Most megacities are expected to grow in the next ten to 15 years, creating new challenges for the USAF and the U.S. military in general. The complexities of megacities, beyond their sheer size, are the consequential differences that make them stand out from other large cities. Dr. Russel Glenn, a senior U.S. Army Training and Doctrine Command advisor, states that "the dense population, complicated structural patterns, and vast infrastructure systems of a megacity makes it a system of systems with no easily identifiable cause and effect; pressure on one point of the system yields reaction and counterpressure elsewhere."[3] Furthermore, the size of a megacity means it connects with regional—and even global—economic and security dynamics in ways that smaller cities do not.

---

[1] James Canton, "The Extreme Future of Megacities," *Significance*, Vol. 8, No. 2, 2011, p. 53.

[2] Richard Dobbs, Sven Smit, Jaana Remes, James Manyika, Charles Roxburgh, and Alejandra Restrepo, *Urban World: Mapping the Economic Power of Cities*, New York: McKinsey Global Institute, 2011, p. 4; United Nations, Department of Economics and Social Affairs, Population Division, *World Urbanization Prospects: The 2014 Revision*, New York, 2015, p. 16.

[3] John Amble and John Spencer, "So You Think the Army Can Avoid Fighting in Megacities," Modern War Institute, May 16, 2017.

Context: Megacities Complicate Military Operations

The size of megacities and the governance challenges they produce can breed areas of lawlessness and fertile grounds for violent nonstate actors to recruit and operate. Coastal megacities are also particularly vulnerable to climatic change, including SLR and extreme weather, rendering them even more unstable.[4] In this environment, these groups can use black markets, shadow governments, illicit economies, and dark networks to gain new members and finance their activities.[5]

At the same time, megacities also raise unique challenges for military operations. Dense masses of people give cover to enemies and their activities. The environment also impairs the required ISR platforms necessary to conduct military operations.[6] Aerial surveillance and close air support can be hindered by tall buildings, narrow streets, and subterranean spaces that protect adversaries.[7]

Historical Trend: Megacities Have Grown in Number, Size, and Importance; These Dense Urban Environments Complicate Military Missions for Advanced Militaries

In 1990, there were ten megacities, serving as home to 153 million people.[8] As of 2016, the United Nations counts 31 megacities with a combined population of 500 million.[9] Cities are also becoming transnational actors and the three trends of urbanization, globalization, and

---

[4] Janani Vivekenanda and Neil Bhatiya, "Coastal Megacities vs. the Sea: Climate and Security in Urban Spaces," in Werrell and Femia, 2017.

[5] Chad Serena and Colin Clarke, "A New Kind of Battlefield Awaits the U.S. Military—Megacities," *The RAND Blog*, April 6, 2016.

[6] Serena and Clarke, 2016.

[7] Austin G. Commons, "Cyber Is the New Air," *Military Review*, Vol. 98, No. 1, January–February 2018.

[8] United Nations, Department of Economics and Social Affairs, Population Division, *A World of Cities*, New York, 2014.

[9] United Nations, Department of Economics and Social Affairs, Population Division, *The World's Cities in 2016—Data Booklet*, New York, 2016, p. 4.

devolution of power to substate entities raise megacities' importance relative to nation-states.[10]

Even advanced militaries face significant challenges when fighting in dense urban environments. Twice during the first Chechen war (1994–1996) and once again from December 1999 until February 2000, the Russian military struggled to seize Grozny from Chechen separatists. Despite Russia's large relative advantage in capabilities, this effort proved difficult because the separatists could use the terrain and local support to their advantage.[11] The Israel Defense Forces faced similar limitations in its three wars against Hamas in Gaza between 2008 and 2014. Ground forces struggled to detect tunnels and fighting in subterranean environments while the Israeli Air Force struggled to stop rocket fire from Gaza, exerting much less pressure on Hamas to reach peace.[12] The U.S. military likewise, endured similar experiences fighting in Baghdad, Sadr City, and Fallujah during the Iraq War. Despite complete technological military dominance, the United States struggled to control Baghdad and Sadr City during much of the U.S. occupation from 2003 to 2011 because of the complexities of urban warfare.[13] The Battles of Fallujah ended up being the bloodiest engagements involving U.S. troops since the Vietnam War.[14] Similarly, the Syrian civil war exposed the many difficulties of conducting ISR against terrorists and insurgents in crowded urban environments.[15] Even in situations where targets could be successfully identified in Islamic State urban strongholds (such as Raqqa), the U.S. military

---

[10] Michele Acuto and Parag Khana, "Nations Are No Longer Driving Globalization—Cities Are," *Quartz*, May 3, 2013.

[11] Serena and Clarke, 2016.

[12] Raphael S. Cohen, David E. Johnson, David E. Thaler, Brenna Allen, Elizabeth M. Bartles, James Cahill, and Shira Efron, *From Cast Lead to Protective Edge: Lessons from Israel's Wars in Gaza*, Santa Monica, Calif.: RAND Corporation, RR-1888-A, 2017.

[13] William Matthews, "Megacity Warfare: Taking Urban Combat to a Whole New Level," webpage, Association of the United States Army, February 15, 2015.

[14] William Head, "The Battles of Al-Fallujah: Urban Warfare and the Growth of Air Power," *Air Power History*, Vol. 60, No. 4, 2013, p. 47.

[15] Serena and Clarke, 2016.

called off some tactical strikes because of collateral damage concerns.[16] Ultimately, megacities present similar, but magnified, complexities to those presented by urban warfare.

Future Projection: Megacities Will Continue to Grow in Number and Size; Violent Nonstate Actors Will Develop in Megacities

According to the United Nations, all current megacities except Tokyo and Osaka will increase in size by 2030 (Figure 7.1).[17] Ten more cities will become megacities by then—Lahore, Pakistan; Hyderabad, India; Bogotá, Colombia; Johannesburg, South Africa; Bangkok, Thailand; Dar es Salaam, Tanzania; Ahmanabad, India; Luanda, Angola; Ho Chi Minh City, Vietnam; and Chengdu, China.[18] Notably, most of the megacities will be in developing nations.[19]

Experts fear that some megacities could turn chaotic and be ruled by gangs and warlords.[20] Current or recent similar situations are found in areas controlled by drug cartels in Colombia and favelas in Brazil.[21] Even if these groups cannot control these cities outright, they can cause chaos and instability, as occurred in Mogadishu and in cities within Haiti, Lebanon, Congo, and Nigeria.[22]

---

[16] Serena and Clarke, 2016.

[17] Department of Economics and Social Affairs, Population Division, 2015, p. 90.

[18] United Nations, Department of Economics and Social Affairs, Population Division, 2016, p. 4.

[19] Canton, 2011, p. 53.

[20] Canton, 2011, pp. 55–56.

[21] Canton, 2011, p. 55.

[22] Canton, 2011, p. 55.

Figure 7.1
The World's Megacities

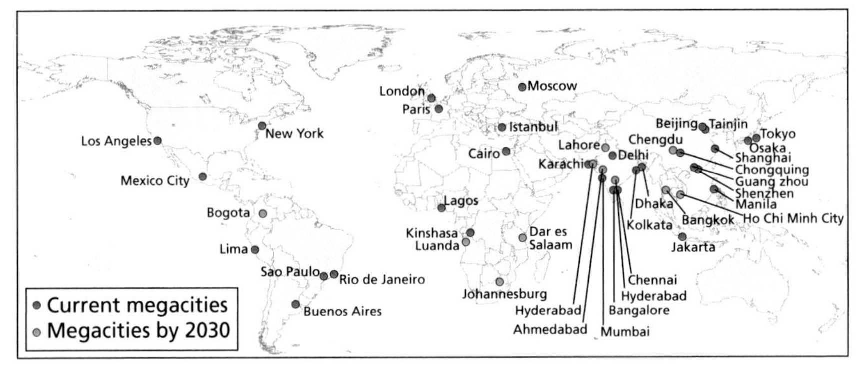

SOURCE: For countries: UNIGIS Geospatial Education Resources, "Countries WGS84," June 1, 2015; for megacities: ESRI, 2017; and United Nations, Department of Economics and Social Affairs, Population Division, 2016, p. 4.

Implications for the U.S. Air Force and the Future of Warfare

According to Army Chief of Staff General Mark Milley, the growth of megacities suggests that the U.S. military will be fighting in dense urban areas in the future.[23] The U.S. Army remains focused on the problem of megacities, but the issue has attracted less attention in the USAF, even though the growth and proliferation of these areas will affect the employment of air and cyber power almost as much as land power.

As with other dense urban environments, megacities greatly decrease the operational advantages earned from air superiority because of the complex physical and human environment.[24] The USAF could be limited in its ability to provide close air support and might not be able to rely on it.[25] Urban environments provide ample opportunities for adversaries to find cover and concealment—whether in buildings

---

[23] Amble and Spencer, 2017.

[24] Commons, 2018.

[25] Commons, 2018.

or subterranean features, such as sewers and tunnels. As the USAF learned in Iraq and Afghanistan and the Israeli Air Force learned in Gaza, conducting strikes within dense urban environments often risks collateral damage, and friendly fire invites international scrutiny.[26]

Dense urban environments also complicate aerial surveillance. Aside from the obvious challenges of being able to collect information through multistory buildings, the availability and density of phone networks, the internet, social media, Google maps, Global Positioning System, encrypted mobile communications, unmanned aerial vehicles, and 3-D printing allow nonstate actors to lower the U.S. military's relative technological and communication advantages when fighting in these environments.[27] The extensive use of these technologies can overload the U.S. military's optical-electrical surveillance platforms, effectively producing an electronic and cyberfog.[28]

Megacities also could exacerbate humanitarian crises. John Amble and Maj. John Spencer of the Modern War Institute state that "the high population density, low quality of life and sanitation conditions, and inability of governments to provide services to vast swaths of megacities also put them at high risk to the rapid spread of diseases like the Zika and Ebola viruses. Megacities are military humanitarian aid and disaster relief operations waiting to happen."[29] The megacity environment will make any HA/DR mission more complex and possibly longer, thus placing greater strain on budgets and human resources.

---

[26] For the problems that the Israeli Air Force encountered during Operation Protective Edge in Gaza in 2014, see Cohen et. al, 2017, pp. 90–96.

[27] Matthews, 2015.

[28] *Cyberfog* refers to a fog network in which data are fragmented and distributed, leaving the information opaque to adversaries. Although this offers the advantage of cyberattack resiliency, the cyberfog approach "presents formidable challenges with respect to data and network management complexity; bandwidth, storage, and battery-power demands; data-reassembly latency; and intermittent connectivity." See Alexander Kott, Ananthram Swami, and Bruce J. West, "The Fog of War in Cyberspace," *Computer*, Vol. 49, No. 11, 2016.

[29] Amble and Spencer, 2017.

CHAPTER EIGHT

Conclusion

This study analyzed how six environmental and geographic trends could affect U.S. national security and the future of warfare. Table 8.1 summarizes the findings from this analysis.

These trends on their own are unlikely to lead to state collapse or interstate conflict, but they are threat multipliers—likely to exacerbate existing problems and fuel instability around the world. This suggests that, despite the 2018 *National Defense Strategy* priorities,[1] the USAF will face continued demand for counterterrorism and stability operations and likely increasing demand for HA/DR missions in the years to come.

Even if these trends will not directly spur great-power conflict, they will shape how the USAF engages in such conflicts. For example, the Ronald Reagan Ballistic Missile Defense Test Site in the Marshall Islands, which plays a key role in detecting missile launches emanating from Asia, could be uninhabitable by 2035.[2] Similarly, Congress noted in the National Defense Authorization Act for Fiscal Year 2018 that "[i]n the Marshall Islands, an Air Force radar installation built on an atoll at a cost of $1,000,000,000 is projected to be underwater within two

[1] U.S. Department of Defense, *Summary of the 2018 National Defense Strategy of the United States of America*, Washington, D.C., 2018.

[2] Curt D. Storlazzi et al., *The Impact of Sea-Level Rise and Climate Change on Department of Defense Installations on Atolls in the Pacific Ocean*, Washington, D.C.: U.S. Department of Defense, RC-2334, February 2018.

Table 8.1
Summary of Findings

Trend	Who Will Fight	How the United States Will Fight	Where the United States Will Fight	Why the United States Will Fight
Rising temperatures		Changes required in extremely hot regions (MENA, South Asia) to lower health risks to service members and to ensure that airbases and aircraft maintain effectiveness	Arid warm regions where ethnic socioeconomic tensions already exist (at risk: MENA, East Africa)	Reduced economic productivity could weaken governments and exacerbate ethnic tensions, prompting more foreign crises
Opening of the Arctic	Arctic powers unlikely to fight over resources or sea routes but economic interests will increase both Russian and Chinese presence in region	U.S. military needs to protect sites from coastal erosion; USAF at disadvantage in terms of Arctic ISR and reliable communications and mapping; harsh environment costly to operate in		Spillover from other conflicts; changing calculus regarding benefits of cooperation versus competition
Sea level rise		USAF could face more demand for HA/DR and counterterrorism missions; rising sea levels could likely affect USAF basing and training	Likely low elevation coastal zones at risk of floods. SLR could spark conflict in South China Sea and affect Chinese terriclaiming efforts.	SLR acts as a threat multiplier exacerbating other causes of conflict (e.g., poverty, political instability and social tensions); displacement and migration could strengthen violent nonstate actors

Table 8.1—Continued

Trend	Who Will Fight	How the United States Will Fight	Where the United States Will Fight	Why the United States Will Fight
Extreme weather events		USAF could face more demand for HA/DR and counterterrorism missions; flooding likely to affect USAF basing and training	Asia Pacific region is most vulnerable—bases will be flooded and overburdened with calls for HA/DR	Could increase spread of disease and lead to migration instigating spikes in violence, civil war, and regime change in developing countries
Water scarcity		USAF could face more demand for HA/DR missions; need to adopt more cost-efficient water and energy uses	Countries unable to mitigate water scarcity in the MENA region, sub-Saharan Africa, and Central and South Asia	Could hinder food security and undermine livelihood in agriculture dependent areas, leading to more water-related intrastate and interstate conflict and unrest
Urbanization and megacities	Terrorist groups, gangs, warlords	Decreased advantage from air superiority; high risk of collateral damage requires more nonlethal, high-precision weapons; HA/DR missions could become more complex and longer, straining budgets and human resources	Megacities in developing countries; those in low-lying areas vulnerable to SLR and extreme weather events	Megacities produce conditions of failed governance and lawlessness, which could require U.S. intervention; also could exacerbate humanitarian crises, including spread of disease

decades."[3] These trends will also shape where and how the USAF trains as the service confronts training spaces at risk of flooding or coastal erosion, storms that damage base infrastructure, and service members coping with health challenges resulting from heat and infectious disease. In sum, environmental and geographical factors will have a bearing on how, where, and why the USAF fights future wars.

We provide some policy recommendations that the USAF could adopt to prepare for and mitigate against the consequences of geographical trends.

Cooperate with other areas of the U.S. government to develop climate data and analysis for USAF needs. The USAF should collaborate with other parts of the U.S. government and the scientific community to ensure improved modeling and analysis suited to its specific needs.[4]

Create the USAF equivalent of the U.S. Navy's Task Force Climate Change. One form of collaboration we have proposed is that the USAF establish a group of experts to anticipate, analyze, and address specific climate change risks to the Air Force Mission—essentially the USAF equivalent of the U.S. Navy's Task Force Climate Change. In addition to members of the USAF, it could be made up with professionals from the National Oceanic and Atmospheric Administration, the National Aeronautics and Space Administration, the U.S. Department of the Interior, international organizations, and other agencies of the U.S. government.

Incorporate risks associated with a changing climate and the proliferation of megacities into planning. The USAF should incorporate future climate scenarios, such as the IPCC scenarios, into planning and war games.[5] Projections on temperature rise, SLR, extreme weather events, and water scarcity should be integrated into regular planning cycles.[6] The USAF should understand the risks associated

---

[3] U.S. House of Representatives Committee on Armed Services, "National Defense Authorization Act for Fiscal Year 2018," H.R. 2810, June 7, 2017.

[4] Messera et al., 2018, p. 41.

[5] Messera et al., 2018, p. 40.

[6] Messera et al., 2018, p. 39.

with these trends and prepare for scenarios in which they fuel civil conflicts or breakdowns in society that could draw the United States into new theaters of warfare—in North Africa, the Middle East, and South Asia. Additionally, planning should incorporate new capacities for humanitarian missions because of possible crises from climate migration.

Develop USAF-specific strategies and capabilities to operate in the Arctic. Although the DoD Arctic strategy does provide some guidance, the USAF was seemingly ignored in the 2013 and 2016 versions of the document.[7] The absence of the USAF from these documents is puzzling because the fastest (and often only) way to respond to a potential crisis in this part of the world is by air. The USAF should develop its own Arctic strategy similar to the Navy and Coast Guard. In a joint manner with the other branches, it should coordinate its defense cooperation with other Arctic nations and invest in capabilities needed to operate in the region, such as reliable communications and ISR to provide real-time monitoring and alerts. The U.S. military also should make certain that it can obtain near real-time maps of the region, particularly because navigable areas are constantly changing.

Assess current USAF infrastructure and plan for future climate risks. The USAF should review which parts of its global infrastructure assets are needed to fulfill its future missions and scrutinize this information against the locations of future climate risks. Depending on the assessments, some sites will need to be closed, moved, or repurposed. Other installations will need to be protected, which could require investments. Installations might need to be created and developed in new locations. The USAF should also seek assistance from other sectors, such as urban planners and architects, that are designing, modifying, and adapting infrastructure and planning spaces to be resilient against more severe and frequent environmental threats.

Augment capabilities to operate in megacities. The USAF and the U.S. military in general both need to develop doctrines, techniques,

---

[7] John L. Conway III, "Toward a U.S. Air Force Arctic Strategy," *Air & Space Power Journal*, Vol. 31, No. 2, 2017, p. 70; DoD, Arctic Strategy, Washington, D.C., November 2013; DoD, 2016.

tactics, procedures, and specialized equipment to operate in the megacity environment. The USAF should ensure it has adequate ISR to operate in dense urban environments, such as the capability to monitor and collect digital communications in megacities.[8] Making that material accessible poses another problem, however: The amount of data that are communicated in megacities would be overwhelming. The USAF, together with intelligence agencies, should therefore develop capabilities to quickly handle "big data"—taking an overflow of information in a dense environment and using it to assemble a comprehensive and actionable intelligence picture in real time.[9] Like other branches of the military, the USAF will also need the capability to operate in a difficult aerial environment of megacities consisting of residential buildings, power lines, and antennas while withstanding rocket-propelled grenades and other fire.[10] Given the concerns about collateral damage, more-precise and more-nonlethal weapons will facilitate the USAF's fight in megacities in the future. Finally, the USAF should possess electronic warfare capabilities that detect, jam, and deceive enemy sensors while allowing the USAF's own communications, positioning, and electrical capabilities to continue functioning.[11]

---

[8] Commons, 2018.

[9] Serena and Clarke, 2016.

[10] Spencer, 2017.

[11] Spencer, 2017.

References

2030 Water Resources Group, *Charting Our Water Future: Economic Frameworks to Inform Decision-Making*, New York, 2009. As of February 25, 2019: https://www.mckinsey.com/-/media/mckinsey/dotcom/client_service/sustainability/pdfs/charting%20our%20water%20future/charting_our_water_future_full_report_.ashx

Acuto, Michele, and Parag Khana, "Nations Are No Longer Driving Globalization—Cities Are," *Quartz*, May 3, 2013. As of May 10, 2018: https://qz.com/80657/the-return-of-the-city-state/

"Agreement on Conservation of Polar Bears," signed by the governments of Canada, Denmark, Norway, the Union of Soviet Socialist Republics and the United States of America, Oslo, November 15, 1973.

Akpodiogaga, Peter, and Odjugo Ovuyovwiroye, "General Overview of Climate Change Impacts in Nigeria," *Journal of Human Ecology*, Vol. 29, No. 1, 2010, pp. 47–55.

Albert, Mathias, Stephan Stetter, Eva Herschinger, and Thomas Teichler, "Conflicts About Water: Securitizations in a Global Context," *Cooperation and Conflict*, Vol. 46, No. 4, 2011, pp. 441–459.

Alexeeva, Olga V., and Frédéric Lasserre, "China and the Arctic," *Arctic Yearbook 2012*, Iceland: Arctic Portal, 2012.

Amble, John, and John Spencer, "So You Think the Army Can Avoid Fighting in Megacities," Modern War Institute, May 16, 2017. As of May 10, 2018: https://mwi.usma.edu/think-army-can-avoid-fighting-megacities/

Anderson, Craig A., Kathryn B. Anderson, Nancy Dorr, Kristina M. DeNeve, and Mindy Flanagan, "Temperature and Aggression," *Advances in Experimental Social Psychology*, Vol. 32, 2000, pp. 63–133.

Arsenault, Chris, "Drought, Expanding Deserts and 'Food for Jihad' Drive Mali's Conflict," Reuters, April 27, 2015.

Associated Press, "Rising Seas Could Threaten $1 Billion Air Force Radar Site," CBS News, October 18, 2016. As of March 4, 2018: https://www.cbsnews.com/news/rising-sea-warnings-air-force-radar-site/

Atkin, Emily, "Inside the U.S. Military's Fight Against Sea Level Rise," *Circa*, January 4, 2017. As of March 4, 2018: https://www.circa.com/story/2017/01/04/politics/inside-the-us-militarys-fight-against-sea-level-rise

Åtland, Kristian, "Mikhail Gorbachev, the Murmansk Initiative, and the Desecuritization of Interstate Relations in the Arctic," *Cooperation and Conflict*, Vol. 43, No. 3, 2008, pp. 289–311.

Beary, Brian, "Race for the Arctic," *CQ Global Researcher*, Vol. 2, August 1, 2008, pp. 213–242.

Bhattacharyya, Arpita, and Michael Werz, *Climate Change, Migration, and Conflict in South Asia*, Washington, D.C.: Center for American Progress, 2012.

Brückner, Markus, and Antonio Ciccone, "Rain and the Democratic Window of Opportunity," *Econometrica*, Vol. 79, No. 3, 2011, pp. 923–947.

Buchanan, Rose Troup, "Global Terrorism Index: Nigerian Fulani Militants Named as Fourth Deadliest Terror Group in World," *Independent*, November 18, 2015.

Budzik, Philip, *Arctic Oil and Natural Gas Potential*, Washington, D.C.: U.S. Energy Information Administration, Office of Integrated Analysis and Forecasting, Oil and Gas Division, 2009.

Buixadé Farré, Albert, et al., "Commercial Arctic Shipping Through the Northeast Passage: Routes, Resources, Governance, Technology, and Infrastructure," *Polar Geography*, Vol. 37, No. 4, 2014, pp. 298–324.

Burek, Peter, Yusuke Satoh, Günther Fischer, Mohammed Taher Kahil, Angelika Scherzer, Syliai Tramberend, Luzma Fabiola Nava, Yoshihide Wada, Stephanie Eisner, Martina Florke, Naota Hanasaki, Piotr Magnuszewski, Bill Cosgrove, and David Wiberg, *Water Futures and Solution: Fast Track Initiative*, Laxenburg, Austria: International Institute for Applied Systems Analysis, Working Paper WP-16-006, 2016.

Burke, Paul J., and Andrew Leigh, "Do Output Contractions Trigger Democratic Change?" *American Economic Journal: Macroeconomics*, Vol. 2, No. 4, 2010, pp. 124–157.

Byers, Michael, "Crises and International Cooperation: An Arctic Case Study," *International Relations*, Vol. 31, No. 4, 2017, pp. 375–402.

Caminade, Cyril, Joanne Turner, Soeren Metelmann, Jenny C. Hesson, Marcus S. C. Blagrove, Tom Solomon, Andrew P. Morse, and Matthew Baylis, "Global Risk Model for Vector-Borne Transmission of Zika Virus Reveals the Role of El Niño 2015," *Proceedings of the National Academy of Sciences*, Vol. 114, No. 1, 2016, pp. 119–124.

Canton, James, "The Extreme Future of Megacities," *Significance*, Vol. 8, No. 2, 2011, pp. 53–56.

Carleton, Tamma A., and Solomon M. Hsiang, "Social and Economic Impacts of Climate," *Science*, Vol. 353, No. 6304, 2016.

Center for Climate and Energy Solutions, "Heat Waves and Climate Change," webpage, undated. As of February 8, 2018:
https://www.c2es.org/content/heat-waves-and-climate-change/

Chaney, Eric, "Revolt on the Nile: Economic Shocks, Religion and Political Influence," *Topics in Middle Eastern and North African Economies*, Vol. 13, 2011.

CNA Military Advisory Board, *National Security and the Threat of Climate Change*, Alexandria, Va.: CNA Corporation, 2007.

———, *National Security and the Accelerating Risks of Climate Change*, Alexandria, Va.: CNA Corporation, 2014.

Coffel, Ethan D., Terence R. Thompson, and Radley M. Horton, "The Impacts of Rising Temperatures on Aircraft Takeoff Performance," *Climatic Change*, Vol. 144, No. 2, 2017, pp. 381–388.

Cohen, Raphael S., Nathan Chandler, Shira Efron, Bryan Frederick, Eugeniu Han, Kurt Klein, Forrest E. Morgan, Ashley L. Rhoades, Howard J. Shatz, and Yuliya Shokh, *The Future of Warfare in 2030: Project Overview and Conclusions*, Santa Monica, Calif.: RAND Corporation, RR-2849/1-AF, 2020. As of May 2020:
https://www.rand.org/pubs/research_reports/RR2849z1.html

Cohen, Raphael S., David E. Johnson, David E. Thaler, Brenna Allen, Elizabeth M. Bartles, James Cahill, and Shira Efron, *From Cast Lead to Protective Edge: Lessons from Israel's Wars in Gaza*, Santa Monica, Calif.: RAND Corporation, RR-1888-A, 2017. As of July 30, 2018:
https://www.rand.org/pubs/research_reports/RR1888.html

Cohen, Raphael S., Eugeniu Han, and Ashley Rhoades, *Geopolitical Trends and the Future of Warfare: The Changing Global Environment and Its Implications for the U.S. Air Force*, Santa Monica, Calif.: RAND Corporation, RR-2849/2, 2020. As of May 2020:
https://www.rand.org/pubs/research_reports/RR2849z2.html

Commander, Naval Installations Command, "Joint Base Anacostia-Bolling—About," webpage, undated. As of April 4, 2018:
http://www.cnic.navy.mil/regions/ndw/installations/jbab/about.html

Commons, Austin G., "Cyber Is the New Air," *Military Review*, Vol. 98, No. 1, January-February 2018, pp. 120–130. As of May 10, 2018:
https://www.hsdl.org/?view&did=806894

Conger, John, "Three Takeaways from Hurricane Michael's Impact on Tyndall Air Force Base," *Climate 101 Security*, blog post, October 19, 2018. As of May 10, 2018:
https://climateandsecurity.org/2018/10/19/three-takeaways-from-hurricane-michaels-impact-on-tyndall-air-force-base/

Conley, Heather A., and Caroline Rohloff, *The New Ice Curtain: Russia's Strategic Reach to the Arctic*, New York: Center for Strategic and International Studies, 2015.

Conway, John L., III, "Toward a U.S. Air Force Arctic Strategy," *Air & Space Power Journal*, Vol. 31, No. 2, 2017, pp. 68–81.

DeBois King, Marcus, "The Weaponization of Water in Syria and Iraq," *Washington Quarterly*, Vol. 38, No. 4, Winter 2016, pp. 153–169.

Defense Intelligence Agency, "About DIA," webpage, undated. As of April 4, 2018:
http://www.dia.mil/About/

Dell, Jan, Susan Tierney, Guido Franco, Richard G. Newell, Rich Richels, John Weyant, and Thomas J. Wilbanks, "Energy Supply and Use," in Jerry M. Melillo, Terese C. Richmond, and Gary W. Yohe, eds., *Climate Change Impacts in the United States: The Third National Climate Assessment*, Washington, D.C.: U.S. Global Change Research Program, 2014, pp. 113–129.

Denezh, Inga, "Russia Plans to Shut Its Northern Sea Route to Foreign Vessels," *Asia Times*, November 22, 2017.

Di Liberto, Tom, "Northwest Passage Clear of Ice Again in 2016," climate.gov, September 16, 2016. As of June 2, 2018:
https://www.climate.gov/news-features/event-tracker/northwest-passage-clear-ice-again-2016

Dillow, Clay, "Russia and China Vie to Beat the U.S. in the Trillion-Dollar Race to Control the Arctic," CNBC, February 6, 2018. As of March 4, 2018:
https://www.cnbc.com/2018/02/06/russia-and-china-battle-us-in-race-to-control-arctic.html

Dobbs, Richard, Sven Smit, Jaana Remes, James Manyika, Charles Roxburgh, and Alejandra Restrepo, *Urban World: Mapping the Economic Power of Cities*, New York: McKinsey Global Institute, 2011.

DoD—*See* U.S. Department of Defense.

Esri, "World Countries," Garmin International, November 30, 2017. As of February 22, 2019:
https://www.arcgis.com/home/item.html?id=d974d9c6bc924ae0a2ffea0a46d71e3d

Eurasia Group, *Opportunities and Challenges for Arctic Oil and Gas Development*, Washington, D.C.: Wilson Center, 2013.

European Academies' Science Advisory Council, Leopoldina—Nationale Akademie der Wissenschaften, "New Data Confirm Increased Frequency of Extreme Weather Events: European National Science Academies Urge Further Action on Climate Change Adaptation," *ScienceDaily*, March 21, 2018. As of April 4, 2018: http://www.sciencedaily.com/releases/2018/03/180321130859.htm

Ferris, Robert, "There's a Scientific Reason Why Hot Weather Has Grounded Planes at Phoenix Airport," CNBC, June 20, 2017. As of February 8, 2018: https://www.cnbc.com/2017/06/20/theres-a-scientific-reason-why-hot-weather-has-grounded-planes-at-phoenix-airport.html

Fetzer, Thiemo, *Can Workfare Programs Moderate Violence? Evidence from India*, London: London School of Economics, working paper, June 2, 2014.

Field, Christopher B., Vicente Barros, Thomas F. Stocker, and Qin Dahe, eds., *Managing the Risks of Extreme Events and Disasters to Advance Climate Change Adaptation: Special Report of the Intergovernmental Panel on Climate Change*, New York: Intergovernmental Panel on Climate Change, 2012.

Flake, Lincoln Edson, "Russia's Security Intentions in a Melting Arctic," *Military and Strategic Affairs*, Vol. 6, No. 1, 2014, pp. 99–116.

Flanders Marine Institute, 2016, Maritime Boundaries Geodatabase, version 10, in conjunction with National Oceanic and Atmospheric Administration, "Exclusive Economic Zones (EEZs)," May 10, 2018. As of February 22, 2019: https://www.arcgis.com/home/item.html?id=5433d0112fc8448e96f61594c900 11c6

Foley, Catherine, *Military Basing and Climate Change*, Washington, D.C.: American Security Project, 2012.

Food and Agriculture Organization of the United Nations, "Maps and Spatial Data," webpage, undated. As of June 15, 2018: http://www.fao.org/nr/water/aquastat/maps/index.stm

Frederick, Bryan, and Nathan Chandler, *Restraint and the Future of Warfare: The Changing Global Environment and Its Implications for the U.S. Air Force*, Santa Monica, Calif.: RAND Corporation, RR-2849/6-AF, 2020. As of May 2020: https://www.rand.org/pubs/research_reports/RR2849z6.html

Fröhlich, Christiane J., Jan Selby, Omar S. Dahi, and Mike Hulme, "Climate Change and the Syrian Civil War Revisited," *Political Geography*, Vol. 60, 2017, pp. 232–244.

Gao, Charlotte, "China Issues Its Arctic Policy," *The Diplomat*, January 26, 2018. As of March 4, 2018: https://thediplomat.com/2018/01/china-issues-its-arctic-policy/

Gardner, Timothy, "Global Powers Strike Deal to Research Before Fishing Arctic Seas," Reuters, November 30, 2017.

Gautier, Donald L., and Thomas E. Moore, *Introduction to the 2008 Circum-Arctic Resource Appraisal (Cara) Professional Paper*, Menlo Park, Calif.: U.S. Geological Survey, 2017.

Ghimire, Ramesh, and Susana Ferreira, "Floods and Armed Conflict," *Environment and Development Economics*, Vol. 21, No. 1, 2016, pp. 23–52.

Gibbs, Ann E., and Bruce M. Richmond, *National Assessment of Shoreline Change—Summary Statistics for Updated Vector Shorelines and Associated Shoreline Change Data for the North Coast of Alaska, U.S.-Canadian Border to Icy Cape*, Washington, D.C.: U.S. Geological Survey, 2017.

Githeko, Andrew, Steve W. Lindsay, Ulisses E. Confalonieri, and Jonathan A. Patz, "Climate Change and Vector-Borne Diseases: A Regional Analysis," *Bulletin of the World Health Organization*, Vol. 78, No. 9, 2000, pp. 1136–1147.

Gleick, Peter H., "Water, Drought, Climate Change, and Conflict in Syria," *Weather, Climate, and Society*, Vol. 6, No. 3, 2014, pp. 331–340.

———, "Water, Security, and Conflict: Violence over Water in 2015," *ScienceBlogs*, February 17, 2016. As of February 25, 2019: https://scienceblogs.com/significantfigures/index.php/2016/02/17/water-security-and-conflict-violence-over-water-in-2015

Great Lakes Integrated Sciences + Assessments, "Extreme Precipitation," webpage, undated. As of June 15, 2018: http://glisa.umich.edu/climate/extreme-precipitation

Hague, William, "The Diplomacy of Climate Change," Russell C. Leffingwell Lecture, Council on Foreign Relations, New York, September 27, 2010.

Hall, John A., Stephen Gill, Jayantha Obeysekera, William Sweet, Kevin Knuuti, and John Marburger, *Regional Sea Level Scenarios for Coastal Risk Management: Managing the Uncertainty of Future Sea Level Change and Extreme Water Levels for Department of Defense Coastal Sites Worldwide*, Alexandria, Va.: Strategic Environmental Research and Development Program, 2016. As of August 27, 2018: http://www.dtic.mil/dtic/tr/fulltext/u2/1013613.pdf

Hallegatte, Stephane, Mook Bangalore, Laura Bonzanigo, Marianne Fay, Tamaro Kane, Ulf Narloch, Julie Rozenberg, David Treguer, and Adrien Vogt-Schilb, *Shock Waves: Managing the Impacts of Climate Change on Poverty*, Washington, D.C.: World Bank, 2015.

Hambling, David, "Does the U.S. Stand a Chance Against Russia's Icebreakers?" *Popular Mechanics*, April 4, 2018. As of August 26, 2018: https://www.popularmechanics.com/military/navy-ships/a19673250/future-icebreakers/

Harari, Mariaflavia, and Eliana La Ferrara, "Conflict, Climate and Cells: A Disaggregated Analysis," *Review of Economics and Statistics*, Vol. 100, No. 4, January 2013.

Hart Crowser, "Environmental Assessment for North Warning System," Washington, D.C.: Department of the Air Force, Air Force Systems Command Electronic Systems Division, November 10, 1986.

Havlík, Petr, Hugo Valin, Mykola Gusti, Erwin Schmid, Nicklas Forsell, Mario Herrero, Nikolay Khabarov, Aline Mosnier, Matthew Cantele, and Michael Obersteiner, *Climate Change Impacts and Mitigation in the Developing World: An Integrated Assessment of the Agriculture and Forestry Sectors*, Washington, D.C.: World Bank Group, Policy Research Working Paper No. WPS 7477, 2015.

Head, William, "The Battles of Al-Fallujah: Urban Warfare and the Growth of Air Power," *Air Power History*, Vol. 60, No. 4, 2013, pp. 32–51.

Herring, David, "Climate Change: Global Temperature Projections," climate.gov, March 6, 2012. As of March 4, 2018:
https://www.climate.gov/news-features/understanding-climate/climate-change-global-temperature-projections

Hidalgo, F. Daniel, Suresh Naidu, Simeon Nichter, and Neal Richardson, "Economic Determinants of Land Invasions," *Review of Economics and Statistics*, Vol. 92, No. 3, 2010, pp. 505–523.

Hijioka, Yasuaki, Erda Lin, and Joy Jacqueline Pereira, "Asia," in Vicente R. Barros and Christopher B. Field, eds., *Climate Change 2014: Impacts, Adaptation, and Vulnerability,* Part B: *Regional Aspects; Working Group II Contribution to the Fifth Assessment Report of the Intergovernmental Panel on Climate Change*, New York: Cambridge University Press, 2014, pp. 1327–1370.

Holland, Andrew, "How Focusing on Climate Security in the Pacific Can Strengthen Alliances: Lessons from the Global Defense Index on Climate Change for the U.S.," in Caitlin E. Werrell and Francesco Femia, eds., *The U.S.–Asia Rebalance, National Security and Climate Change*, Washington, D.C.: Center for American Progress, November 2015. As of April 4, 2018:
https://climateandsecurity.files.wordpress.com/2017/02/asiapacific_holland_ch4.pdf

Horton, Alex, "U.S. Jets Intercept Pair of Russian Bombers off Alaskan Coast," *Washington Post*, May 12, 2018. As of March 7, 2018:
https://www.washingtonpost.com/news/checkpoint/wp/2018/05/12/u-s-jets-intercept-pair-of-russian-bombers-off-alaskan-coast/?utm_term=.01993644f345

Hotez, Peter, Kristy O. Murray, and Pierre Buekens, "The Gulf Coast: A New American Underbelly of Tropical Diseases and Poverty," *PLoS Neglected Tropical Diseases*, Vol. 8, No. 5, 2014.

Hsiang, Solomon M., Marshall Burke, and Edward Miguel, "Quantifying the Influence of Climate on Human Conflict," *Science*, Vol. 341, No. 6151, 2013.

Hsiang, Solomon M., and Amir S. Jina, *The Causal Effect of Environmental Catastrophe on Long-Run Economic Growth: Evidence from 6,700 Cyclones*, Cambridge, Mass.: National Bureau of Economic Research, NBER Working Paper, No. w20352, 2014.

Hughes, Zachariah, "Climate Accelerating Erosion at U.S. Radar Facilities in Arctic," *Barents Observer*, July 6, 2016.

Im, Eun-Soon, Jeremy S. Pal, and Elfatih A. B. Eltahir, "Deadly Heat Waves Projected in the Densely Populated Agricultural Regions of South Asia," *Science Advances*, Vol. 3, No. 8, 2017.

Intergovernmental Panel on Climate Change, "Guidance Note for Lead Authors of the IPCC Fifth Assessment Report on Consistent Treatment of Uncertainties," IPCC Cross-Working Group Meeting on Consistent Treatment of Uncertainties, Jasper Ridge, Calif., July 6–7, 2010. As of June 20, 2018: https://wg1.ipcc.ch/SR/documents/ar5_uncertainty-guidance-note.pdf

Intergovernmental Panel on Climate Change Working Group II, *Climate Change 2014: Impacts, Adaptation and Vulnerability,* Part A: *Global and Sectoral Aspects,* New York: Cambridge University Press, 2014. As of February 8, 2018: https://www.ipcc.ch/site/assets/uploads/2018/02/WGIIAR5-PartA_FINAL.pdf

International Displacement Migration Centre, *Global Estimates 2015: People Displaced by Disasters*, Geneva, 2015. As of April 4, 2018: http://www.internal-displacement.org/publications/2015/global-estimates-2015-people-displaced-by-disasters/

International Organization of Migration, Department of Operations and Emergencies, *Dimensions of Crisis on Migration in Somalia*, Geneva, working paper, February 2014.

IPCC—*See* Intergovernmental Panel on Climate Change.

Ives, Mike, "North Korea Aside, Guam Faces Another Threat: Climate Change," *New York Times*, August 11, 2017. As of March 4, 2018: https://www.nytimes.com/2017/08/11/world/asia/guam-north-korea-climate-change.html

Jackman, Albert, "The Nature of Military Geography," *The Professional Geographer*, Vol. 14, 1962, pp. 7–12.

Jacob, Brian, Lars Lefgren, and Enrico Moretti, "The Dynamics of Criminal Behavior Evidence from Weather Shocks," *Journal of Human Resources*, Vol. 42, No. 3, 2007, pp. 489–527.

Jiménez Cisneros, Blanca E., Taikan Oki, Nigel W. Arnell, Gerardo Benito, J. Graham Cogley, Petra Doll, Tong Jiang, and Shadrack S. Mwakalila, "Freshwater Resources," in Christopher B. Field and Vicente Barros, eds., *Climate Change 2014: Impacts, Adaptation, and Vulnerability,* Part A: *Global and Sectoral Aspects; Working Group II Contribution to the Fifth Assessment Report of the Intergovernmental Panel on Climate Change*, New York: Cambridge University Press, 2014, pp. 229–269.

Kelley, Colin P., Shahrzad Mohtadi, Mark A. Cane, Richard Seager, and Yochanan Kushnir, "Climate Change in the Fertile Crescent and Implications of the Recent Syrian Drought," *Proceedings of the National Academy of Sciences*, Vol. 112, No. 11, 2015, pp. 3241–3246.

Knies, Jochen, Patricia Cabedo-Sanz, Simon T. Belt, Soma Baranwal, Susanne Fietz, and Antoni Rosell-Melé, "The Emergence of Modern Sea Ice Cover in the Arctic Ocean," *Nature Communications*, Vol. 5, 2014.

Kott, Alexander, Ananthram Swami, and Bruce J. West, "The Fog of War in Cyberspace," *Computer*, Vol. 49, No. 11, 2016, pp. 84–87.

Kube, Courtney, "Russian Tu-95 Bombers Fly Near Alaskan Coast, Again," NBC News, April 19, 2017. As of June 9, 2018:
https://www.nbcnews.com/news/us-news/russian-bombers-fly-near-alaskan-coast-again-n748611

Lindsey, Rebecca, "Climate Change: Global Sea Level," climate.gov, August 1, 2018. As of March 4, 2018:
https://www.climate.gov/news-features/understanding-climate/climate-change-global-sea-level

Lindsey, Rebecca, and LuAnn Dahlman, "Climate Change: Global Temperature," climate.gov, September 11, 2017. As of February 8, 2018:
https://www.climate.gov/news-features/understanding-climate/climate-change-global-temperature

Lostumbo, Michael J., Michael J. McNerney, Eric Peltz, Derek Eaton, David R. Frelinger, Victoria Greenfield, John Halliday, Patrick Mills, Bruce R. Nardulli, Stacie L. Pettyjohn, Jerry M. Sollinger, and Stephen M. Worman, *Overseas Basing of US Military Forces: An Assessment of Relative Costs and Strategic Benefits*, Santa Monica, Calif.: RAND Corporation, RR-201-OSD, 2013. As of February 22, 2019:
https://www.rand.org/pubs/research_reports/RR201.html

Luo, Tianyi, Robert Samuel Young, and Paul Reig, "Aqueduct Projected Water Stress Country Rankings," Washington, D.C., World Resources Institute, data set, August 2015.

Masters, Jeff, "Super Typhoon Haiyan Storm Surge Survey Finds High Water Marks 46 Feet High," *Weather Underground*, May 8, 2014. As of April 4, 2018:
https://www.wunderground.com/blog/JeffMasters/super-typhoon-haiyan-storm-surge-survey-finds-high-water-marks-46-feet.html

———, "Historic Heat Wave Sweeps Asia, the Middle East and Europe," *Weather Underground*, June 6, 2017. As of February 8, 2018:
https://www.wunderground.com/cat6/historic-heat-wave-sweeps-asia-middle-east-and-europe

Mathis, Jeremy, "Fishing in the Arctic?" *NOAA Research News*, September 8, 2017. As of March 7, 2018:
https://research.noaa.gov/article/ArtMID/587/ArticleID/27/Fishing-in-the-Arctic

Matthews, William, "Megacity Warfare: Taking Urban Combat to a Whole New Level," webpage, Association of the United States Army, February 15, 2015. As of May 10, 2018:
https://www.ausa.org/articles/
megacity-warfare-taking-urban-combat-whole-new-level

Maystadt, Jean-Francois, Olivier Ecker, and Athur Mabiso, *Extreme Weather and Civil War in Somalia: Does Drought Fuel Conflict Through Livestock Price Shocks?* Washington, D.C.: International Food Policy Research Institute, IFPRI Discussion Paper 01243, 2013.

McGrath, Matt, "China Joins Arctic Council but a Decision on the EU Is Deferred," BBC News, May 15, 2013. As of March 7, 2018:
https://www.bbc.com/news/science-environment-22527822

McLeman, Robert, "Migration and Displacement in a Changing Climate," in Werrell and Femia, 2017, pp. 100–107.

———, "Migration and Displacement Risks Due to Mean Sea-Level Rise," *Bulletin of the Atomic Scientists*, Vol. 74, No. 3, 2018, pp. 148–154.

McMahon, Patrice C., "Cooperation Rules: Insights on Water and Conflict from International Relations," in Jean Axelron Cahan, ed., *Water Security in the Middle East: Essays in Scientific and Social Cooperation*, London: Anthem Press, 2017, pp. 19–38.

McMichael, Anthony J., and Andrew Haines, "Global Climate Change: The Potential Effects on Health," *British Medical Journal Clinical Research*, Vol. 315, No. 7111, 1997, pp. 805–809.

Mekonnen, Mesfin M., and Arjen Y. Hoekstra, "Four Billion People Facing Severe Water Scarcity," *Science Advances*, Vol. 2, No. 2, February 2016.

Messera, Heather, Ronald Keys, John Castellaw, Robert Parker, Ann C. Phillips, Jonathan White, and Gerald Galloway, *Military Expert Panel Report: Sea Level Rise and the U.S. Military's Mission*, 2nd ed., Washington, D.C.: Center for Climate & Security, 2018.

Miguel, Edward, Shanker Satyanath, and Ernest Sergenti, "Economic Shocks and Civil Conflict: An Instrumental Variables Approach," *Journal of Political Economy*, Vol. 112, No. 4, 2004, pp. 725–753.

Mizokami, Kyle, "Why Is North Korea So Fixated on Guam?" *Popular Mechanics*, August 9, 2017. As of March 4, 2018:
https://www.popularmechanics.com/military/a27687/north-korea-guam/

Mollman, Steve, "It's Typhoon Season in the South China Sea—and China's Fake Islands Could Be Washed Away," *Quartz*, August 1, 2016. As of March 4, 2018: https://qz.com/745511/international-law-isnt-the-most-powerful-threat-to-chinas-artificial-islands-in-the-south-china-sea-nature-is/

Mooney, Chris, "Warming of the Arctic Is 'Unprecedented over the Last 1,500 Years,' Scientists Say," *Washington Post*, December 12, 2017.

Morgan, Forrest E., and Raphael S. Cohen, *Military Trends and the Future of Warfare: The Changing Global Environment and Its Implications for the U.S. Air Force*, Santa Monica, Calif.: RAND Corporation, RR-2849/3-AF, 2020. As of May 2020:
https://www.rand.org/pubs/research_reports/RR2849z3.html

National Aeronautics and Space Administration, collections search, undated. As of August 6, 2018:
https://lpdaac.usgs.gov/dataset_discovery/measures/measures_products_table

———, "The Impact of Climate Change on Natural Disasters," March 3, 2005. As of June 15, 2018:
https://earthobservatory.nasa.gov/Features/RisingCost/rising_cost5.php

National Aeronautics and Space Administration Goddard Institute for Space Studies, "Long-Term Warming Trend Continued in 2017: NASA, NOAA," webpage, January 18, 2018. As of February 8, 2018:
https://www.giss.nasa.gov/research/news/20180118/

National Centers for Environmental Information, "Billion-Dollar Weather and Climate Disasters," undated. As of February 5, 2018:
https://www.ncdc.noaa.gov/billions/

———, "Arctic Graticule," January 18, 2018. As of February 22, 2019:
https://www.arcgis.com/home/item.html?id=5d6866e3771c4d4397d9ad59f556ada5

National Geospatial-Intelligence Agency, Arctic Sea Routes, January 15, 2016. As of February 22, 2019:
https://www.arcgis.com/home/item.html?id=9d3669ec9e0a44209b7ad8d9aabb766a

Nawyn, Stephanie J., "Migration in the Global South: Exploring New Theoretical Territory," *International Journal of Sociology*, Vol. 46, No. 2, 2016, pp. 81–84.

Naylor, Hugh, "An Epic Middle East Heat Wave Could Be Global Warming's Hellish Curtain-Raiser," *Washington Post*, August 10, 2016. As of February 8, 2018: https://www.washingtonpost.com/world/middle_east/an-epic-middle-east-heat-wave-could-be-global-warmings-hellish-curtain-raiser/2016/08/09/c8c717d4-5992-11e6-8b48-0cb344221131_story.html?utm_term=.c410f24513e6

Nilsen, Thomas, "Russia Opens New Rescue Base on Northern Sea Route," *Independent Barents Observer*, via *Eye on the Arctic*, January 12, 2018.

NOAA National Centers for Environmental Information, "Global Climate Report—Annual 2016," webpage, undated. As of February 8, 2018:
https://www.ncdc.noaa.gov/sotc/global/201613

O'Connor, Tom, "China and Russia May Take Over the Top of the World with New 'Polar Silk Road' Through the Arctic," *Newsweek*, January, 26, 2018. As of March 4, 2018:
http://www.newsweek.com/
china-russia-may-take-over-top-world-new-polar-silk-road-792490

Office of the Under Secretary of Defense for Acquisition, Technology, and Logistics, *Department of Defense Climate-Related Risk to DoD Infrastructure Initial Vulnerability Assessment Survey (SLVAS) Report*, Washington, D.C.: U.S. Department of Defense, January 2018. As of February 8, 2018:
https://climateandsecurity.files.wordpress.com/2018/01/tab-b-slvas-report-1-24-2018.pdf

Ohio State University, "Climate Change Threatens Drinking Water, As Rising Sea Penetrates Coastal Aquifers," *ScienceDaily*, November 7, 2007.

Oliver-Smith, Anthony, "Sea Level Rise and the Vulnerability of Coastal Peoples: Responding to the Local Challenges of Global Climate Change in the 21st Century," UNU-EHS InterSecTions, Vol. 7, 2009.

Pachauri Rajendra K., et al., *Climate Change 2014 Synthesis Report*, Geneva: Intergovernmental Panel on Climate Change, 2014.

Pacific Institute, "Water Conflict Chronology List," webpage, 2018. As of July 30, 2018:
http://www.worldwater.org/conflict/list/

Palanisamy, Hindumathi, Anny Cazenave, Benoit Meyssignac, Laurent Soudarin, Guy Wöppelmann, and Melanie Becker, "Regional Sea Level Variability, Total Relative Sea Level Rise and Its Impacts on Islands and Coastal Zones of Indian Ocean over the Last Sixty Years," *Global and Planetary Change*, Vol. 116, 2014, pp. 54–67.

Panama National Environmental Authority, *Enhancing Resilience to Climate Change and Climate Variability in the Central Pacific Region of Panama*, April 2013. As of February 8, 2018:
http://www.miambiente.gob.pa/images/stories/BibliotecaVirtualImg/
CambioClimatico/Propuesta_de_Panama_al_FA_en.pdf

Parham, Paul, "Hard Evidence: Will Climate Change Affect the Spread of Tropical Diseases?" *The Conversation*, February 17, 2015. As of February 8, 2018:
http://theconversation.com/
hard-evidence-will-climate-change-affect-the-spread-of-tropical-diseases-37566

Patterson, Tom, and Nathaniel Vaughn Kelso, vector and raster map data, Natural Earth, undated. As of June 12, 2018:
https://www.naturalearthdata.com

Pellerin, Cheryl, "Recovery Effort Takes on Great Energy, Task Force Commander Says," DoD News, November 19, 2013. As of April 4, 2018:
http://archive.defense.gov/news/newsarticle.aspx?id=121177

Pezard, Stephanie, Abbie Tingstad, Kristin Van Abel, and Scott Stephenson, *Maintaining Arctic Cooperation with Russia: Planning for Regional Change in the Far North*, Santa Monica, Calif.: RAND Corporation, RR-1731-RC, 2017. As of February 22, 2019:
https://www.rand.org/pubs/research_reports/RR1731.html

Philipps, Dave, "Tyndall Air Force Base a 'Complete Loss' Amid Questions About Stealth Fighters," *New York Times*, October 11, 2018.

Rasul, Golam, "Managing the Food, Water, and Energy Nexus for Achieving the Sustainable Development Goals in South Asia," *Environmental Development*, Vol. 18, April 2016, pp. 14–25.

Revkin, Andrew, "Trump's Defense Secretary Cites Climate Change as National Security Challenge," ProPublica, March 14, 2017. As of February 4, 2018:
https://www.propublica.org/article/
trumps-defense-secretary-cites-climate-change-national-security-challenge

Ribot, Jesse, "Vulnerability Does Not Fall from the Sky: Toward Multiscale, Pro-Poor Climate Policy," in Robin Mearns and Andrew Norton, eds., *Social Dimensions of Climate Change: Equity and Vulnerability in a Warming World*, Washington, D.C.: World Bank, 2010, pp. 47–74.

Samaras, Constantine, "U.S. Military Basing Considerations During a Rebalance to Asia: Maintaining Capabilities Under Climate Change Impact," in Caitlin E. Werrell and Francesco Femia, eds., *The U.S.–Asia Rebalance, National Security and Climate Change*, Washington, D.C.: Center for American Progress, November 2015, pp. 34–35.

Samenow, Jason, "Two Middle East Locations Hit 129 Degrees, Hottest Ever in Eastern Hemisphere, Maybe the World," *Washington Post*, July 22, 2016. As of February 8, 2018:
https://www.washingtonpost.com/news/capital-weather-gang/wp/2016/07/22/two-middle-east-locations-hit-129-degrees-hottest-ever-in-eastern-hemisphere-maybe-the-world/?utm_term=.25a931ca59a0

———, "Iranian City Soars to Record 129 Degrees: Near Hottest on Earth in Modern Measurements," *Washington Post*, June 29, 2017. As of February 8, 2018:
https://www.washingtonpost.com/news/capital-weather-gang/wp/2017/06/29/
iran-city-soars-to-record-of-129-degrees-near-hottest-ever-reliably-measured-on-earth/?utm_term=.a9cd21b43b75

Sandler, Todd, "The Analytical Study of Terrorism Taking Stock," *Journal of Peace Research*, Vol. 51, No. 2, 2014, pp. 257–271.

Sarsons, Heather, "Rainfall and Conflict: A Cautionary Tale," *Journal of Development Economics*, Vol. 115, No. C, 2015, pp. 62–72.

Schär, Christoph, "Climate Extremes: The Worst Heat Waves to Come," *Nature Climate Change*, Vol. 6, No. 2, 2016, pp. 128–129.

Schleussner, Carl-Friedrich, Jonathan F. Donges, Reik V. Donner, and Hans Joachim Schellnhuber, "Armed-Conflict Risks Enhanced by Climate-Related Disasters in Ethnically Fractionalized Countries," *Proceedings of the National Academy of Sciences*, Vol. 113, No. 33, 2016, pp. 9216–9221.

Serena, Chad, and Colin Clarke, "A New Kind of Battlefield Awaits the U.S. Military—Megacities," *The RAND Blog*, April 6, 2016. As of February 25, 2019: https://www.rand.org/blog/2016/04/a-new-kind-of-battlefield-awaits-the-us-military-megacities.html

Shatz, Howard J., and Nathan Chandler, *Global Economic Trends and the Future of Warfare: The Changing Global Environment and Its Implications for the U.S. Air Force*, Santa Monica, Calif.: RAND Corporation, RR-2849/4, 2020. As of May 2020:
https://www.rand.org/pubs/research_reports/RR2849z4.html

Sims, Alexandra, "India's Roads Melt as Record-Breaking Heat Wave Continues," *Independent*, May 23, 2016. As of February 8, 2018:
http://www.independent.co.uk/news/world/asia/india-s-roads-melt-as-record-breaking-heat-wave-continues-a7044146.html

Smith, Laurence C., and Scott R. Stephenson, "New Trans-Arctic Shipping Routes Navigable by Midcentury," *Proceedings of the National Academy of Sciences*, Vol. 110, No. 13, 2013, pp. E1191–E1195.

Solow, Andrew R., "Global Warming: A Call for Peace on Climate and Conflict," *Nature*, Vol. 497, 2013.

Spanger-Siegfried, Erika, Kristina Dahl, Astrid Caldas, and Shana Udvardy, "The U.S. Military on the Front Lines of Rising Seas: Exposure to Coastal Flooding at Joint Base Anacostia-Bolling and Washington Navy Yard, Washington, District of Columbia," Cambridge, Mass.: Union of Concerned Scientists, fact sheet, July 2016. As of April 4, 2018:
https://www.ucsusa.org/sites/default/files/attach/2016/07/us-military-on-front-lines-of-rising-seas_all-materials.pdf

Staalesen, Atle, "Russian Legislators Ban Foreign Shipments of Oil, Natural Gas and Coal Along Northern Sea Route," *Barents Observer*, December 26, 2017. As of June 2, 2018:
https://thebarentsobserver.com/en/arctic/2017/12/russian-legislators-ban-foreign-shipments-oil-natural-gas-and-coal-along-northern-sea

Steinbruner, John D., Paul C. Stern, and Jo L. Husbands, eds., *Climate and Social Stress: Implications for Security Analysis*, Washington, D.C.: Committee on Assessing the Impacts of Climate Change on Social and Political Stresses, Board on Environmental Change and Society, Division of Behavioral and Social Sciences and Education, National Research Council, 2013.

Sternberg, Troy, "Chinese Drought, Bread and the Arab Spring," *Applied Geography*, Vol. 34, No. 4, May 2012, pp. 519–524.

Stocker, T. F., et al., *Climate Change 2013: The Physical Science Basis*, Geneva: Intergovernmental Panel on Climate Change, 2013.

Storlazzi, Curt D., et al., *The Impact of Sea-Level Rise and Climate Change on Department of Defense Installations on Atolls in the Pacific Ocean*, Washington, D.C.: U.S. Department of Defense, RC-2334, February 2018. As of June 15, 2018: https://www.serdp-estcp.org/Program-Areas/Resource-Conservation-and-Resiliency/Infrastructure-Resiliency/Vulnerability-and-Impact-Assessment/RC-2334/

Thompson, John, "Putin Plays Mr. Nice Guy at Russian Arctic Forum," *Arctic Deeply*, March 30, 2017.

Tritten, Travis, "When Disaster Strikes, U.S. Military Assets Often Key to Relief Efforts," *Stars and Stripes*, November 16, 2013.

UNIGIS Geospatial Education Resources, "Countries WGS84," June 1, 2015. As of February 22, 2019:
https://hub.arcgis.com/datasets/a21fdb46d23e4ef896f31475217cbb08_1

United Nations, Department of Economics and Social Affairs, Population Division, *A World of Cities*, New York, 2014.

———, *World Urbanization Prospects: The 2014 Revision*, New York, 2015.

———, *The World's Cities in 2016—Data Booklet*, New York, 2016. As of February 18, 2019:
http://www.un.org/en/development/desa/population/publications/pdf/urbanization/the_worlds_cities_in_2016_data_booklet.pdf

United Nations Office for Disaster Risk Reduction, "Easing Impact of Drought on the Panama Canal," June 28, 2016. As of April 4, 2018:
https://www.unisdr.org/archive/49408

U.S. Department of Defense, *FY 2012 Climate Change Adaptation Roadmap*, Washington, D.C., September 18, 2012. As of April 4, 2018:
http://www.dodworkshops.org/Appendix_A_-_DoD_Climate_Change_Adaption_Roadmap_20120918.pdf

———, *Arctic Strategy*, Washington, D.C., November 2013.

———, *Report to Congress on Strategy to Protect United States National Security Interests in the Arctic Region*, Washington, D.C., 2016.

———, "Military Installations, Ranges, and Training Areas," data set, January 18, 2017. As of April 4, 2018:
https://catalog.data.gov/dataset/military-installations-ranges-and-training-areas

———, *Summary of the 2018 National Defense Strategy of the United States of America*, Washington, D.C., 2018.

U.S. Department of Energy, *U.S. Energy Sector Vulnerabilities to Climate Change and Extreme Weather*, Washington, D.C., July 2013. As of February 8, 2018: https://www.energy.gov/sites/prod/files/2013/07/f2/20130716-Energy%20Sector%20Vulnerabilities%20Report.pdf

U.S. Department of State, "Global Water Security 2012," Intelligence Community Assessment ICA 2012-08, February 2, 2012.

U.S. Government Accountability Office, *DOD Can Improve Infrastructure Planning and Processes to Better Account for Potential Impacts*, Washington, D.C., report to congressional requesters, May 2014.

"USGS National Elevation Dataset (NED)," data.gov, undated. As of August 10, 2018:
https://catalog.data.gov/dataset/usgs-national-elevation-dataset-ned

U.S. House of Representatives Committee on Armed Services, "National Defense Authorization Act for Fiscal Year 2018," H.R. 2810, June 7, 2017. As of June 15, 2018:
https://www.congress.gov/bill/115th-congress/house-bill/2810

"U.S. Military Bases in Kuwait," MilitaryBases.com, undated. As of February 8, 2018:
https://militarybases.com/overseas/kuwait/

Van Lange, Paul A. M., Maria I. Rinderu, and Brad J. Bushman, "Aggression and Violence Around the World: A Model of Climate, Aggression, and Self-Control in Humans (CLASH)," *Behavioral and Brain Sciences*, Vol. 40, 2017.

Vivekenanda, Janani, and Neil Bhatiya, "Coastal Megacities vs. the Sea: Climate and Security in Urban Spaces," in Werrell and Femia, 2017.

VornDick, Wilson T., "Thanks Climate Change: Sea-Level Rise Could End South China Sea Spat," *The Diplomat*, November 8, 2012. As of March 4, 2018: https://thediplomat.com/2012/11/can-climate-change-wash-away-south-china-sea-dispute/?allpages=yes

———, "China's Island Building + Climate Change: Bad News," *Real Clear Defense*, March 9, 2015a. As of March 4, 2018:
https://www.realcleardefense.com/articles/2015/03/10/chinese_island_reclamation_the_climate_change_challenge_107722-2.html

———, "Terriclaims: The New Geopolitical Reality in the South China Sea," Asia Maritime Transparency Initiative, April 8, 2015b. As of June 30, 2018: https://amti.csis.org/terriclaims-the-new-geopolitical-reality-in-the-south-china-sea/

Wallin, Matthew, "U.S. Military Bases and Facilities in the Middle East," Washington, D.C.: American Security Project, fact sheet, June 2018. As of April 30, 2019:
https://www.americansecurityproject.org/wp-content/uploads/2018/07/Ref-0213-US-Military-Bases-and-Facilities-Middle-East.pdf

Walsh, Kevin J. E., John L. McBride, Philip J. Klotzbach, Sethurathinam Balachandran, Suzana J. Camargo, Greg Holland, Thomas R. Knutson, James P. Kossin, Tsz-cheung Lee, Adam Sobel, and Masato Sugi, "Tropical Cyclones and Climate Change," *WIREs Climate Change*, Vol. 7, No. 1, January–February 2016, pp. 65–89.

Warner, Koko, Charles Ehrhart, Alex de Sherbinin, Susana Adamo, and Tricia Chai-Onn, *In Search of Shelter: Mapping the Effects of Climate Change on Human Migration and Displacement*, Bonn, Germany: United Nations University, CARE, and CIESIN-Columbia University and in close collaboration with the European Commission "Environmental Change and Forced Migration Scenarios Project," the UNHCR, and the World Bank, 2009.

Werrell, Caitlin E., and Francesco Femia, eds., *The Arab Spring and Climate Change: A Climate and Security Correlations Series*, Washington, D.C.: Center for American Progress, 2013a.

———, eds., *Climate Change Before and After the Arab Awakening: The Cases of Syria and Libya*, Washington, D.C.: Center for Climate and Security, February 2013b.

———, "General Keys: The Military Thinks Climate Change Is Serious," Washington, D.C.: Center for Climate and Security, July 7, 2016.

———, eds., *Epicenters of Climate and Security: The New Geostrategic Landscape of the Anthropocene*, Washington, D.C.: Center for Climate and Security, June 2017.

White, Chris, "Understanding Water Scarcity: Definitions and Measurements," Global Water Forum, May 7, 2012. As of June 25, 2018:
http://www.globalwaterforum.org/2012/05/07/understanding-water-scarcity-definitions-and-measurements/

Willis, Henry H., David G. Groves, Jeanne S. Ringel, Zhimin Mao, Shira Efron, and Michele Abbott, *Developing the Pardee RAND Food-Energy-Water Security Index*, Santa Monica, Calif.: RAND Corporation, TL-165-RC, 2016. As of February 25, 2019:
https://www.rand.org/pubs/tools/TL165.html

Wodon, Quentin, Andrea Liverani, George Joseph, and Nathalie Bougnoux, *Climate Change and Migration: Evidence from the Middle East and North Africa*, Washington, D.C.: World Bank, 2014.

Wong, Andrew, "China: We Are a 'Near-Arctic State' and We Want a 'Polar Silk Road,'" CNBC, February 14, 2018. As of March 7, 2018:
https://www.cnbc.com/2018/02/14/china-we-are-a-near-arctic-state-and-we-want-a-polar-silk-road.html

World Bank, "High and Dry: Climate Change, Water, and the Economy," Washington, D.C., 2016.

World Economic Forum, *Global Risks Report 2018*, 13th ed., Geneva, 2018.

World Food Program, *Syria Arab Republic Joint Rapid Food Security Needs Assessment (JRFSNA)*, Rome: United Nations Food and Agriculture Organization, 2012.

World Health Organization, "Vector-Borne Diseases," October 31, 2017. As of April 4, 2018:
http://www.who.int/mediacentre/factsheets/fs387/en/

———, "Drinking-Water," fact sheet, February 7, 2018. As of February 25, 2019:
https://www.who.int/news-room/fact-sheets/detail/drinking-water

World Water Assessment Program, *Managing Water Under Uncertainty and Risk*, Paris: United Nations Educational, Science, and Cultural Organization, World Water Development Report 4, 2012. As of June 18, 2018:
http://unesdoc.unesco.org/images/0021/002156/215644e.pdf

———, *Nature-Based Solutions for Water*, Paris: United Nations Educational, Science, and Cultural Organization, World Water Development Report, 2018.

Wright, Brian, and Carlo Cafiero, "Grain Reserves and Food Security in the Middle East and North Africa," *Food Security*, Vol. 3, Suppl. 1, February 2011, pp. 61–76.

Wright, Tom, "Crush of Refugees Inflames Karachi," *Wall Street Journal*, August 26, 2010. As of March 4, 2018:
https://www.wsj.com/articles/SB10001424052748704540904575451512217585010

Zhou, Laura, "Slowly but Surely, China Is Carving a Foothold Through the Arctic," *South China Morning Post*, January 26, 2018.

Zivin, Joshua Graff, and Matthew Neidell, "Temperature and the Allocation of Time: Implications for Climate Change," *Journal of Labor Economics*, Vol. 32, No. 1, 2014, pp. 1–26.